UNIVERSITY OF STRATHCLYDE

30125 00531675 6

Books are to be returned on or before
the last date below.

3 JUN 1997

MANAGEMENT OF PROCESS INDUSTRY WASTE

An Introduction

MANAGEMENT OF PROCESS INDUSTRY WASTE

An Introduction

Edited by Richard Bahu,
Barry Crittenden and John O'Hara

INSTITUTION OF CHEMICAL ENGINEERS

The information in this book is given in good faith and belief in its accuracy, but does not imply the acceptance of any legal liability or responsibility whatsoever, by the Institution, the steering group or by the editors, for the consequences of its use or misuse in any particular circumstances.

All rights reserved. No part of this publication may be reproduced, stored in a retrieval system, or transmitted, in any form or by any means, electronic, mechanical, photocopying, recording or otherwise, without the prior permission of the publisher.

**Published by
Institution of Chemical Engineers,
Davis Building,
165–189 Railway Terrace,
Rugby, Warwickshire CV21 3HQ, UK.**

© 1997 Institution of Chemical Engineers
A Registered Charity

ISBN 0 85295 324 0

Photographs on pages 6, 62, 86, 100, 180 and 200 are reproduced by courtesy of AEA Technology, Culham.

Printed in the United Kingdom by Bookcraft Ltd, Bath.

ACKNOWLEDGEMENTS

Preparation of this book was commissioned by the Institution of Chemical Engineers under the guidance of a steering committee. The Institution gratefully acknowledges that the text is the result of contributions from many individuals associated with the steering group over several years. The Institution thanks the following individuals for their contribution to the steering group work:

L. Barber
M.W. Finch (Zeneca)
J.C. Holden (Merseyside Waste Disposal Unit)
A.Q. Khan (South Yorkshire Hazardous Waste Unit)
R. Mehta (British Gas)
F.J. Owens
D.R. Parsons (Boots)
H.T. Wilson (Best Global Ltd)
R. Wolstenholme
I. Woodhouse (Albright and Wilson)

The final text was edited and significantly expanded by Richard Bahu with the editorial help of Professor Barry Crittenden of the School of Chemical Engineering at the University of Bath. In addition, John O'Hara of Denton Hall, London, made a significant contribution to the sections on legal aspects of waste management.

CONTENTS

		PAGE
ACKNOWLEDGEMENTS		iii
1.	**INTRODUCTION**	1
2.	**FRAMEWORK FOR WASTE MANAGEMENT**	7
2.1	INTRODUCTION	7
2.2	GENERAL PRINCIPLES	7
2.3	WASTE LAW AND REGULATORY AUTHORITIES	12
2.4	PROCESS AND WASTE CHARACTERIZATION	13
2.5	WASTE MINIMIZATION	14
2.6	ON-SITE WASTE TREATMENT	16
2.7	OFF-SITE WASTE DISPOSAL	18
2.8	SAFETY AND HEALTH ISSUES	19
3.	**WASTE LAW AND REGULATORY AUTHORITIES**	21
3.1	INTRODUCTION	21
3.2	DEFINITIONS OF WASTE	24
3.3	LAND USE PLANNING	31
3.4	POLLUTION CONTROL AND WASTE REGULATION	31
3.5	CARRIAGE OF WASTE	44
3.6	DISCHARGES TO SEWER	45
3.7	DISCHARGES TO CONTROLLED WATERS	46
3.8	RADIOACTIVE SUBSTANCES	46
3.9	COMMON LAW	47
3.10	REGULATORY AUTHORITIES	48
3.11	EC LEGISLATION	48
3.12	US ENVIRONMENTAL LAW	57
4.	**PROCESS AND WASTE CHARACTERIZATION**	63
4.1	INTRODUCTION	63
4.2	PROCESS CHARACTERIZATION	63
4.3	WASTE ASSESSMENT AND SURVEY	65
4.4	CHARACTERIZATION AND ANALYSIS	78
4.5	MONITORING	78
4.6	WASTE DATABASE	83
4.7	MARKING AND LABELLING	84

5.	**WASTE MINIMIZATION**	87
5.1	INTRODUCTION	87
5.2	ORGANIZATION	89
5.3	BASIC APPROACH	90
5.4	TECHNIQUES	92
5.5	PRIORITIES AND TARGETS	97
5.6	REVIEWS, AUDITS AND CORRECTIONS	98
5.7	THE CLUB APPROACH	98
6.	**ON-SITE WASTE TREATMENT**	101
6.1	INTRODUCTION	101
6.2	SELECTION OF TREATMENT METHODS	102
6.3	GASEOUS TREATMENT TECHNOLOGIES	105
6.4	LIQUID TREATMENT TECHNOLOGIES	124
6.5	SOLIDS TREATMENT TECHNOLOGIES	134
6.6	INCINERATION	135
7.	**OFF-SITE WASTE DISPOSAL**	157
7.1	INTRODUCTION	157
7.2	DISCHARGES TO AIR	157
7.3	DISCHARGES TO AQUEOUS ENVIRONMENT	159
7.4	DISCHARGES TO LAND	159
7.5	LANDFILL	162
7.6	LIQUID/SOLID WASTE CONTRACTORS	164
7.7	TRANSPORT OF WASTE	168
7.8	TRANSFRONTIER SHIPMENT OF WASTE	173
8.	**SAFETY AND HEALTH ISSUES**	175
8.1	INTRODUCTION	175
8.2	LEGAL FRAMEWORK	176
8.3	CONFINED SPACES	177
8.4	DEEP TANKS	177
8.5	COMBUSTIBLE MATERIALS	178
8.6	VOLATILE ORGANIC COMPOUNDS	178
8.7	ANAEROBIC DIGESTERS	178
8.8	BIOLOGICAL HAZARDS	178
8.9	TOXICITY	179
8.10	CORROSIVITY	179
8.11	CARCINOGENICITY	179

APPENDICES
1 — USEFUL CONTACTS   181
2 — SOFTWARE   183
3 — BIBLIOGRAPHY   185
4 — GLOSSARY OF WASTE DISPOSAL TERMS   187
5 — SPECIAL WASTE CONSIGNMENT NOTE SYSTEM   191
6 — WASTE MANAGEMENT PAPERS (HMSO)   197

INDEX   201

1. INTRODUCTION

WHY HAS THIS BOOK BEEN WRITTEN?
Waste materials are inevitably generated during the manufacture of any product. Figure 1.1 on page 2 gives an outline of where wastes originate and how they flow in virtually all processes. The management of these wastes is of paramount importance to the process industries. This book was written in response to various requests to draw together the diverse strands of the whole subject into a single introductory text.

WHO IS THE BOOK AIMED AT?
The book is written for young engineers and those new to the topic who need a comprehensive overview of the field, the key issues and where to go for further information and assistance. It covers the major elements involved in the management of process industry waste — waste law and regulatory authorities, process and waste characterization, waste minimization, on-site waste treatment, off-site waste disposal and health and safety issues.

WHAT IS THE SCOPE OF THE BOOK?
The process industries include chemical, pharmaceutical, biotechnology, food and drink, oil and gas, power generation and minerals sectors. This list is not exhaustive and any process which takes a raw material or feedstock and applies a physical, chemical or biochemical transformation to yield a new product is included. The term 'waste' is often limited in interpretation as 'solid waste'. This book uses 'waste' in its widest sense to cover gaseous, fluid and solid materials. Indeed, 'waste' equipment and plant — that is, decommissioning, reuse and disposal of redundant hardware — falls within the scope. In particular, decommissioning projects can generate significant quantities of waste from cleaning and dismantling operations. Waste management in the nuclear industry and of radioactive materials are excluded, since these require special procedures.

It is difficult to write on the subject of waste without concentrating on one national system and this book concentrates primarily on the UK. It does, however, cover European Community issues and highlights the US system. At

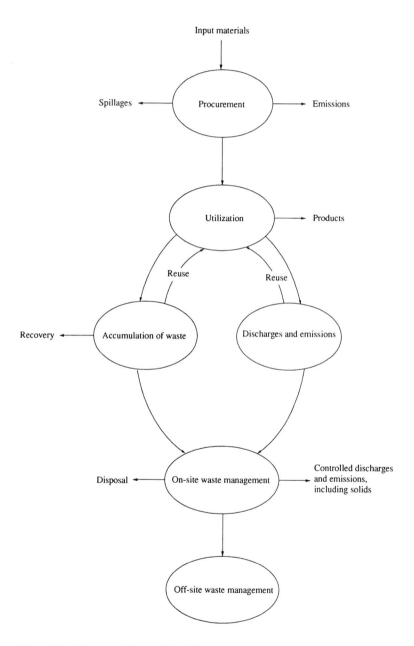

Figure 1.1 Conceptual waste flow diagram.

the present time, legal and regulatory systems are in the process of major evolution. The book aims to give as complete a picture as possible of the topic as it currently exists and to highlight proposals for future planned changes.

WHAT IS THE BACKGROUND TO WASTE MANAGEMENT?

Mounting public concern over the environment has led to the introduction of environmental legislation. This should be seen as an opportunity rather than a threat, as the benefits are numerous. Waste could be the result of an inefficient process. Reducing waste by improving efficiency will also maximize profits. The resulting improvement in the local environment could have significant public relations value. Waste management procedures are often driven by legislation but should be thought of as company-friendly. The terms 'hazardous, toxic and poisonous waste' are not well defined and there is no international consensus. Current thinking on waste management in Europe is driven by a hierarchy of waste management practices set out by the European Commission, namely:

- prevention;
- minimization;
- recycling;
- disposal.

Clearly, it is better to prevent or avoid generating waste in the first place. In practice, complete elimination of waste in a process may be unrealistic and minimization must be the next step. Where a process is left with a residual amount of waste, opportunities to recycle either within the process, on the same site or at another location are better than treatment and disposal.

Against the background of this waste management hierarchy, the scale of the industrial waste disposal business in the UK in 1995 can be gauged from estimates of some 245 million tonnes of controlled wastes and some 2.9 million tonnes of special wastes[1]. Of the special wastes some 85% went direct to landfill, 4% to sea, 4% was incinerated and 7% was treated physically and/or chemically before being landfilled[1]. Disposal to sea is now banned for many wastes and is being phased out for most others in the next few years, even if the environmental arguments favour sea disposal; note the debacle in 1996 over how to dispose of Shell's redundant North Sea storage platform Brent Spar.

Figure 1.2, page 4, gives a pictorial overview of the principal routes for waste treatment and disposal:

A — direct discharge of liquid waste to sewer;
B — *in-situ* treatment of liquid waste prior to discharge to sewer;
C — direct disposal of waste via vehicular transport to landfill;
D — solidification of waste prior to landfill;

E — treatment in a central treatment facility yielding liquid waste to sewer and solid waste to landfill;

F — incineration of waste yielding liquid waste to sewer and solid waste to landfill;

G — direct disposal of waste via vehicular transport and ocean deposit.

In addition, gaseous wastes arise from the production plant, treatment facilities, incineration and fugitive releases.

The general principles of waste management apply to liquid, gaseous and solid wastes. Good waste management practices are essential; each site needs a comprehensive procedure for the management of specific waste streams generated at that site. Written procedures are developed and the responsibilities of individuals defined. Periodically, these procedures and responsibilities are reviewed and, if required, modified. To this end an individual manager is designated to have overall responsibility for the implementation, assessment and updating of the waste management system.

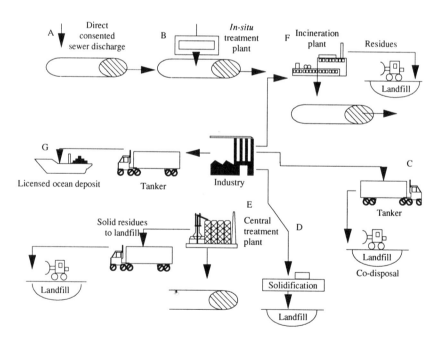

Figure 1.2 Principal industrial effluent routes. Note: discharges may also be made to rivers and coastal estuaries subject to consent. (Holmes[1].)

The manager has overall responsibility for the training and motivation of personnel at all levels to adopt sound waste management practices. The manager is also responsible for the on-site management of all streams generated by the site. This includes ensuring compliance with statutory controls and company guidelines and identifying opportunities for prevention, minimization, recycling, recovery or reuse.

In order to monitor and characterize all wastes, a centralized documentation system is required. Features of such a system might include:
- a waste database to enable waste to be traced from the moment it is generated at the production plant to the moment it is finally disposed of;
- adequate accounting to allow quantification of all waste streams and their disposal costs.

The generation and disposal of wastes is unlikely to receive adequate attention unless an appropriate and relevant costing structure is in place. Waste disposal costs should be charged visibly and directly to the product whose manufacture generates the waste. It is therefore important for waste disposal to be a cost centre in its own right. Without these measures, experience shows that there is little incentive to reduce generation or manage wastes properly.

REFERENCES IN CHAPTER 1
1. Holmes, J.R., 1995, *The United Kingdom Waste Management Industry* (Institute of Waste Management, UK).

2. FRAMEWORK FOR WASTE MANAGEMENT

2.1 INTRODUCTION

This chapter sets out the framework in which the management of waste is undertaken in the process industries. After outlining some general considerations on company management systems, good management practices, public relations and standards, subsequent sections relate to each of the following chapters, where more details can be found. Table 2.1 on page 8 gives an overview of Chapter 2.

2.2 GENERAL PRINCIPLES

The management of process industry waste forms an integral part of a company's environmental management system and ultimately of the overall management system. There is a growing trend to integrate safety, health and environmental (SHE) management in industrial companies.

The discipline of management can be generalized as data gathering, planning, monitoring and implementing all of the necessary steps to achieve a given objective within certain constraints. In the case of the management of waste, the objective is cost-effective handling, minimization, treatment and disposal of waste under the key constraint of legislation and its enforcement. Advice on good management of solid waste is contained in *Industrial Waste Management: A CEFIC Approach to the Issue*[1].

Waste is an emotive topic and the reaction 'Not In My Back Yard' (NIMBY) is all too common. Public concerns and pressure groups are affecting company policy as in particular the chemical industry strives to improve its image. The UK Chemical Industries Association (CIA)[2] and the European Chemical Industry Council (CEFIC) promote such programmes as 'Responsible Care'. The other well-used term is 'Sustainable Development'[3], which is now widespread in UK Government, Confederation of British Industry (CBI), European Union (EU) and United Nations (UN) actions. It is defined as 'Development that meets the needs of the present without compromising the ability of future generations to meet their own needs'. This fundamental premise should be applied to all waste management processes; they all need to become more sustainable.

TABLE 2.1
Overview of Chapter 2

2.2, page 7 General principles	The basis for a management system is set out including the roles of the Waste Manager and the Plant Manager. Public concerns and relations are noted. The need for an appropriate and relevant costing structure is highlighted.
2.3, page 12 Waste law and regulatory authorities (see Chapter 3)	Companies should ensure that they are aware of all relevant legislation covering liquid, gaseous and solid waste streams. The UK Environmental Protection Act 1990 enshrines waste minimization and the best practicable environmental option for disposal. It is essential to establish good working relationships with regulatory authorities, which in England and Wales is the Environment Agency and in Scotland is the Scottish Environment Protection Agency.
2.4, page 13 Process and waste characterization (see Chapter 4)	Waste arisings should be properly surveyed, characterized, monitored and labelled. A database should be set up to provide useful and reliable information and to allow trends to be monitored.
2.5, page 14 Waste minimization (see Chapter 5)	The need to minimize waste generation is a priority consideration, including opportunities to recycle or reuse waste streams
2.6, page 16 On-site waste treatment (see Chapter 6)	Pretreatment and treatment options should be assessed to achieve cost-effective control of waste on-site. Wastes awaiting disposal should be stored in an environmentally sound manner in designated areas which are properly supervised and regularly inspected. Where relevant, a site licence for storage must be obtained.
2.7, page 18 Off-site waste disposal (see Chapter 7)	Discharges to air, aqueous environment and disposal of solid and mixed wastes is often the ultimate end-point for material that cannot be economically utilized. The off-site disposal may include prior on-site or off-site treatment, if this is an attractive option.
2.8, page 19 Safety and health issues (see Chapter 8)	The handling, treatment and discharge of wastes can create a unique combination of safety and health issues

Effective communications and transparency of information on waste between the company, the local community and the public at large is now seen as essential in maintaining trust and credibility. Publication of annual corporate environmental reports is on the increase. CEFIC published reporting guidelines in 1993[4]. Some companies have established community advisory panels covering SHE issues.

An interesting approach to waste management is to make any waste generated highly visible within the company. One danger of a highly automated and efficient waste management system is that it can disappear into the background activities of the plant/site. By keeping waste highly visible, it ensures a sustained programme of action. Although moving waste storage facilities from the back of a site to a location adjacent to the front reception area may be frowned upon, placing transparent sections in waste pipelines — or viewing windows in treatment units — is more acceptable.

The introduction of the Environmental Management System (EMS) standard BS 7750 in 1992 is also influencing the practice of waste management[5]. It mirrors the quality standard BS 5750. The international equivalents of BS 7750 are ISO 14000 and 14001.

A voluntary EU initiative was launched in 1995 called 'Eco-Management and Audit Scheme' (EMAS). To register, a company must:
(1) draw up a corporate environmental policy;
(2) carry out a site environmental review;
(3) introduce an action programme and EMS;
(4) carry out regular environmental audits to check the effectiveness of the EMS;
(5) set new objectives based on the audits;
(6) prepare an environmental statement on the audited site;
(7) submit all of the above for independent verification;
(8) send the environmental statement to the national authority.

Of these actions, independent verification is clearly of most concern to companies in adding costs and duplication of effort. It seems likely that some such scheme may become mandatory in the future, so companies which participate at an early stage may be able to influence its shape and gain some advantage.

Insurance can play an important role in waste management by providing protection against civil liabilities, including those which involve clean-up costs. It is not possible, however, to cover criminal penalties or the consequences of conviction, although the legal costs associated with proceedings can be included. Insurance is operated under 'utmost good faith' where the insured and insurer have a duty to disclose all relevant facts. As insurance policies operate annually, there can be difficulties when a claim arises over damage which may have taken some time to emerge. So there may be advantages to 'claims made' policies for waste liabilities. However, insurers are excluding long-term pollution cover resulting from 'normal' operations and are offering cover for 'sudden and accidental' damage. The CIA has produced an 'Environmental Impairment Liability Insurance Facility' (CEILIC) policy, which is tailored to environmental problems. It is site-specific and covers some degree of on-site clean-up. Lennon[6] has recently reviewed environmental insurance.

In any new process, the implications of waste must be considered at all stages from conceptual design through to detailed design/optimization and pilot studies. In general, the same good waste management techniques and procedures apply to either a new or an existing process. The main differences are that:
- real data can be obtained from the existing process, whereas assumptions may have to be made in the new process;
- the new process may offer greater scope for good waste management as there are likely to be fewer existing constraints;
- if the new process is not replacing an existing old process, it may simply face greater scrutiny as it will be perceived as 'adding' to the environmental load.

It is best to discuss disposal options with regulatory bodies or contractors at as early a stage as possible in any new process. This helps to define the critical components.

The presentation of product to customers must also be considered in a new process. Is the proposed packaging recyclable or readily disposed of? If so it may provide a competitive edge; if not, it may present customers with a problem.

The generation and disposal of wastes is unlikely to receive adequate attention unless an appropriate costing structure is in place. The guiding principle is that of 'the polluter pays'. Costs of disposing of a waste stream should be charged, visibly and directly, to the product whose manufacture generates the waste. A pollution index has been proposed based on the amount of waste generated per unit mass of product. It is, therefore, important for waste disposal to be a cost centre in its own right. The costs should form part of plant standards and be subjected to regular review.

But there are concerns over the use of cost-benefit analysis to support environmental decisions. Some costs can be difficult to segregate and extract from the company financial system, and benefits need to be related to risk. Clearly, the highest risk to the business is that the plant operation could be stopped for a serious infringement. The highest risk to the environment could be a major pollution incident with significant loss of life. There is no consistent basis on which environmental expenditure is calculated and performance assessed; each company has developed its own approach. Kolluru[7] has described a framework for relating risk assessment and financial returns. Moilanen and

Martin[8] have developed a new comprehensive approach for environmental investments based on Expected Monetary Value which addresses the problems of quantifying risk and future liabilities.

Each operating site needs comprehensive waste management policies and procedures where responsibilities are well defined. In particular, each individual's legal responsibilities must be understood clearly. A specific waste manager is designated to have overall responsibility for the implementation, assessment and updating of the waste management system. The waste manager produces an annual report summarizing waste production, treatment and disposal routes, costs, any potential problem areas and future trends. Individual plant managers are responsible for the wastes generated by their particular plant.

The waste manager has overall responsibility for ensuring compliance with statutory controls and company guidelines including:
- motivating, training and assisting all plant personnel in the proper management of waste;
- setting up a documentation system for on-site waste control, ensuring the provision of the necessary information and documentation associated with the treatment, storage and disposal of individual wastes;
- auditing performance and costs of on-site management of all streams (gaseous, liquid and solid) generated by the site including treatment and, where appropriate, handling, storage, transportation and contractors;
- keeping in contact with regulatory authorities, such as the Environment Agency;
- ensuring effective communications with the local community and other interested parties;
- maintaining up-to-date knowledge of legal changes which could affect waste management operations and amending procedures accordingly;
- keeping abreast of technical developments and identifying opportunities for recycling, recovery, reuse, improved treatment or pretreatment such as segregating/blending wastes in order to improve handling or to facilitate disposal;
- storing of wastes awaiting removal from the site;
- approval of the final disposal route.

The plant manager is responsible locally for:
- ensuring viable and cost-effective waste minimization and improved pretreatment-treatment and treatment practices are adopted in consultation with the waste manager;
- providing details of the waste streams and any envisaged changes to the waste manager;
- checking and signing-off that the waste conforms in composition and physical properties to the registered information;

- monitoring performance of on-plant abatement systems, discharges to site treatment systems and segregation or mixing of wastes as specified, and ensuring that waste is placed in appropriate containers and correctly labelled;
- reporting of accidents and incidents related to waste materials and streams;
- ensuring an awareness of projects or other developments which may affect waste streams from the plant.

2.3 WASTE LAW AND REGULATORY AUTHORITIES

Much of waste management practice is driven by legislation (see Chapter 3). The law covering waste management has been significantly changed in the UK by the Environmental Protection Act 1990 (EPA 1990), particularly Part I covering pollution control and Part II covering waste on land.

Part I of EPA 1990 establishes, for the most potentially polluting and technologically complex industrial processes, a system of Integrated Pollution Control (IPC). Under this system the operators of prescribed processes must obtain an authorization which contains legally enforceable conditions for achieving certain objectives. Where the process is likely to involve releases into more than one environmental medium, the operator must have regard to the best practicable environmental option (BPEO) so as to minimize pollution of the environment as a whole. As a result, discharges to air, aqueous environment and land are now linked. Discharges of waste to any single medium cannot be considered in isolation. The other important aspect is the obligation to use best available techniques not entailing excessive cost (BATNEEC).

Part II of the EPA 1990 introduces two important aspects — firstly, a statutory 'duty of care' for waste. It is imposed on, and links, all persons dealing with the waste from its creation to its disposal, 'cradle to grave', so that it is always managed responsibly.

Secondly, Part II introduces a new waste management licensing system superseding the waste disposal regime in the UK Control of Pollution Act 1974. The new system has a wider scope than the old regime, including the keeping, treatment and disposal of waste. It includes the requirement that the holder of a waste management licence must be a 'fit and proper person' in terms of technical competence, financial security and so on.

The other major Acts which are relevant to waste management in the UK include The Control of Pollution (Amendment) Act 1989 which is important as it requires carriers of waste to be registered. The Water Industry Act 1991 and the Water Resources Act 1991 are concerned respectively with discharges to sewers and controlled waters. The Radioactive Substances Act 1993 relates to the holding and disposal of radioactive waste. The other legal elements are the common law and increasingly, European Community law. The most notable European Union (EU) resolution established a hierarchy with waste minimization as the most preferable option, followed by recycling/reuse, with the optimization of waste treatment and disposal at the bottom.

The regulatory authorities have undergone a complete re-organization into the Environment Agency (EA) in England and Wales (which combines the former powers and duties of Her Majesty's Inspectorate of Pollution (HMIP), the National Rivers Authority (NRA) and the Waste Regulation Authorities (WRAs)). In Scotland, the Scottish Environment Protection Agency has been formed from Her Majesty's Industrial Pollution Inspectorate (HMIPI) and the Regional Purification Boards (RPB's).

2.4   PROCESS AND WASTE CHARACTERIZATION

It is the responsibility of the waste generator to provide information on the nature of wastes to others concerned, particularly the authorities and those carrying out treatment, transport and disposal. For each plant, and collectively for each site, an 'adequate' knowledge of the composition and characteristics of the wastes generated must be available (see Chapter 4). The term 'adequate' is discussed below.

The majority of waste streams are likely to be complex mixtures and in practice it is often impracticable, or even impossible, to determine their precise compositions and characteristics. Nevertheless, a broad assessment of the physical, chemical and toxicological properties of each waste arising is necessary for three main reasons:
- to take appropriate measures to protect the health of people handling the waste and the environment, from its generation to its final disposal;
- to select the most appropriate method of disposal;
- to classify the waste in accordance with regulatory requirements, in particular for notification and transportation of solid wastes and to obtain authorization for producing processes under the requirements of EPA 1990.

Some wastes can be characterized simply from a knowledge of the process from which they originate. In particular, certain processes and substances are called 'prescribed' and require specific procedures as set out in EPA 1990. In general, a waste assessment or survey is undertaken and some analytical examination is required to provide adequate information for the survey using either in-house facilities or, more frequently, external specialist analytical services.

Consider the following points when deciding what is necessary to characterize any waste 'adequately':
- data relating to handling and transport — physical state, viscosity, vapour

pressure and so on;
- possible harmful physical properties — flammability, explosivity and so on;
- ecotoxicological and toxicological properties;
- possible nuisance problems — for example, odour;
- data related to incineration — calorific value, halogen and inorganics content and so on;
- data related to landfilling — solubility, hydrolysis, biodegradability, flash point, odour and so on;
- the need to determine special data — such as polychlorinated biphenyls (PCBs) or heavy metals content — or to consider specific sampling procedures, particularly if waste is not homogeneous.

A key point is that segregation of waste streams may facilitate characterization. Take care, however, when mixing waste streams to fully consider any physical, chemical and toxicological interactions.

Having decided on the characterization, frequency of analysis and so on, careful consideration should be given to ensuring that relevant information is passed in an appropriate format to all who need it. In order to monitor and characterize all waste arisings, a centralized documentation system is required. Features of such a system might include:
- a waste database to enable waste to be traced from the moment it is generated at the production plant to the moment it is finally disposed of;
- adequate accounting to allow quantification of all waste streams and their disposal costs.

A database is necessary to provide reliable information to management and the regulatory authorities. It allows trends to be monitored. It can assist decision-making in relation to plant modifications or new plant. Implementation of this type of system makes good business sense because it generates a proper interest in what is being discarded, stimulates the need to investigate potential savings and encourages the development of better waste management.

If the waste is packaged then appropriate marking and labelling must be used, and there are regulations covering the transport of dangerous goods which may apply.

2.5 WASTE MINIMIZATION

Industrial processes inevitably generate a certain amount of residues. For economic as well as environmental reasons, it is in the waste generator's interest to minimize residues through process optimization or redesign in developing cleaner processes. Beyond such development, opportunities to utilize residues in a productive way must be explored through, for example, recycle or recovery as feedstock, for energy production or use. Only if all else fails to be viable should these residues become waste streams to be treated and disposed of. Another important aspect of waste minimization is taking back spent product or surplus stock from customers for reprocessing or repacking. Note that recycling can be viable if it costs less than the disposal options. Recycling is likely to increase as levies are introduced on landfilling and packaging waste.

Waste minimization implies activity at a number of levels in the organization:
- company management — a quantified programme for waste minimization should be established, publicized and monitored. Investment proposals and projects are only approved where evidence is provided that proper consideration has been given to waste management and that waste arisings have been reduced to the practical minimum;
- research and engineering — the need to minimize waste generation is a priority consideration at all stages of research, development and engineering;
- production — the quantity of wastes generated should be monitored and mechanisms established to highlight any accidental or long-term deviations. Staff at all levels should be kept aware of the need to minimize the amount of waste generated in accordance with the objectives of the programme. Segregating wastes may facilitate recycling (but this is normally only possible for materials of relatively simple and consistent composition).

There is a check-list of points to be considered in waste minimization:
- review the points at which waste can arise (from both new and existing installations), including unwanted products of a chemical reaction;
- critically examine processes generating the most significant waste streams and consider, for example, alternative raw materials, reagents or solvents;
- review any process where the achieved conversion efficiency deviates significantly from the theoretical;
- periodically check the amount of waste generated against the standards and objectives;
- develop and implement programmes for regular inspection and maintenance of plants to limit mechanical failures which may be sources of accidental generation of waste;
- when establishing start-up and shut-down procedures, bear in mind the need to minimize the quantity of waste and off-grade product generated during these transient steps.

There is a check-list of points to be considered in the recycle or recovery of residues as raw material or source of raw materials:
- it is often possible to find the potential to recycle a certain waste within the same company, especially in large companies;

- some governmental agencies or trade associations have set up 'residue exchange systems' and regularly publish lists of a wide range of potentially recyclable materials offered by residue generators. Although most of the business carried out by these clearing houses is at the regional level, international deals are not unknown.

Finally, a check-list of points to be considered in the recovery of residues for use as a fuel:
- check the suitability of dealing with the waste by combustion with the regulatory authorities;
- some residues have a calorific value high enough to be used for steam or power generation in industrial boilers, usually in addition to other conventional fuels;
- it is also possible to use the calorific value of certain wastes to support combustion of other less combustible materials. This can result in significant energy saving;
- some high-calorific residues, which cannot be used in conventional boilers because of the acidity of their combustion gases, can be added to fuel oil firing in certain special circumstances such as in cement kilns;
- after combustion, most waste leaves a residue and gaseous emissions, the disposal of which require proper management.

2.6 ON-SITE WASTE TREATMENT

As a general principle, waste treatment methods are preferred to waste storage methods as a long-term solution. They may be undertaken either on or off the site. The simplest waste treatment step is to discharge it safely direct to the environment; this is considered under off-site disposal in Section 2.7 (page 18) and in more detail in Chapter 7.

Interim storage on-site must be considered to facilitate on-site treatment or prior to off-site disposal. The following points cover general guidance on interim storage:
- ensure that storage areas are suitably sited and of adequate capacity. They should be capable of containing any spillages and leakages, or be equipped with secure drainage and ventilation systems. Deal with spillages or leaking containers promptly in a prescribed manner;
- store waste inside buildings or designated compounds. It is easier to check and safeguard;
- position storage areas to allow easy access for deposit and collection;
- label all wastes — use of a bar code system, both for identification and control of waste movements, could be considered;

FRAMEWORK FOR WASTE MANAGEMENT

- follow written procedures for acceptance, storage and removal of wastes. Make them known to all personnel involved with operating a storage facility;
- restrict access to storage areas to designated personnel and make suitable security arrangements;
- give clear instructions on how to deal with emergencies — for example, fire, spillage, contamination of personnel. Supply emergency equipment;
- take precautions to guard against inappropriate mixing of wastes. Segregate incompatible materials;
- specify the types of storage to be used for particular wastes;
- regulate the use of skips and monitor them regularly to ensure that no unauthorized materials are deposited. Incidents have occurred when materials have been placed in the wrong skip. In addition, a small quantity of special waste would lead to the whole load being so designated;
- supervise storage areas properly and inspect them regularly;
- do not use damaged or poor quality drums — they can lead to accidents;
- compactors can play a useful role in decreasing the volume of certain wastes — for example, packaging materials. They must be used with great care, however, and regularly inspected and maintained.

The CIA Guidelines for Safe Warehousing[9] and the UK Health and Safety Executive (HSE) Guidance Note CS 17[10], gives advice on the storage of packaged dangerous goods.

A treatment or disposal method is selected that is appropriate, considering the characteristics of the waste stream, and is in line with the best practicable environmental option (BPEO). More information is provided in Chapter 6. As a general rule, the waste generator is responsible for the selection of the disposal method — unless the use of a specific means of waste disposal is obligatory or forbidden by legislation. The waste treatment and/or disposal method must give adequate protection to the public and the environment and satisfy legislative and other local requirements.

There is a check-list of points to be considered:
- think about ways of treating the waste at source to make it easier, safer or cheaper to dispose of. Methods which can be employed include physical treatment (dewatering, compaction, absorption, etc) or chemical treatment (neutralization, oxidation, reduction, precipitation, etc) or biochemical treatment (anaerobic digesters, activated sludge equipment, etc);
- decide if it is prudent to mix wastes before disposal. Wastes which individually are harmless may become dangerous, or difficult to handle and dispose of, if brought into contact with each other. On the other hand, sometimes mixing waste can make it easier to handle and/or reduce problems of disposal;
- consider the characteristics of the waste, which may include:

— the physical and chemical characteristics;
— the presence of any biologically hazardous components;
— the concentration of contaminants;
— the quantity and frequency of waste generation;
— whether the waste will be in bulk or in drums;
— the likely environmental effects, including the mobility and the persistence of pollutants;

- weigh up the available treatment or disposal methods, taking into account the types of waste for which they would be suitable, the conditions necessary for efficient operations, and the distance any solid wastes would be transported. If technically and economically justified, preference is given to methods enabling the disposal of the wastes as close as possible to the place where they were generated;
- for any single waste stream the range of disposal methods available is usually restricted for ecological, technical or economic reasons. In addition, local circumstances and legislative requirements rule out some methods immediately. For some waste no second choice of disposal method is available;
- provide clear instructions for identifying solid waste in storage and in transit, and check that they are being carried out;
- provide for regular collection of solid waste from routine operations;
- review disposal methods regularly, and be aware of legislative changes affecting disposal plans.

2.7 OFF-SITE WASTE DISPOSAL

Landfill is by far the most popular method of off-site disposal in the UK, due to its relatively low cost and the UK Government's position on market-driven solutions to environmental issues. The introduction of a landfill levy, however, will encourage greater use of off-site recycling and treatment technologies.

Landfill is a sound solution provided high standards of design, operation and long-term management are maintained to protect groundwater and air pollution, including landfill gas control and utilization.

There are many cases where waste materials cannot be disposed of on the site where they are generated. It is then necessary to use the services of specialist contractors for the transport, treatment and disposal of wastes. Specific guidance on the use of waste contractors is given in CIA Waste Management Guideline No.1 entitled 'The Use of Waste Disposal Contractors'. Although this deals mainly with transported waste, the information is also relevant where, for example, use is made of a municipal waste water treatment system connected by pipeline. Part II of EPA 1990 'duty of care' is particularly relevant here. See Section 2.3 (page 12) and Chapter 3 for more details.

When external transport or disposal contractors are employed, the company must be satisfied that they can deal with waste materials safely, effectively and legally, and confirm that waste consignments reach the specified final disposal site and are disposed of in the agreed environmentally safe manner.

Select contractors carefully on the basis of their ability to handle and process wastes, using methods and practices which provide adequate safety to people and the environment and which comply with all applicable laws and regulations. Furthermore, waste generators should protect their interest by documenting the selection of contractors, formalizing all arrangements by written agreements, obtaining assurances that waste is reaching the final disposal site and that the materials are dealt with in the agreed manner.

The responsibility for the waste generator's co-ordination and control of waste contractors is assigned to a manager at the location where a given waste material is being generated. The nominated manager must be a person of technical competence ensuring that the following are done by formal audits (see Section 7.6, page 164):

- the disposal site is chosen as close as possible to the site of generation of the waste in order to reduce the movements to a minimum when this is ecologically, technically and economically justified;
- information is collected on proposed contractors and proposed arrangements evaluated;
- proposed arrangements are recommended to management for acceptance;
- relevant environmental, safety, health and technical data are transferred to the contractor prior to the first shipment of the material in question;
- an arrangement is formalized by contract with the contractor prior to the first shipment;
- assurances are obtained that all waste is transferred, treated or disposed of at named locations according to contractual obligations and provisions of the relevant laws and regulations;
- necessary steps are taken to comply with international transport conventions and/or national regulations.

2.8 SAFETY AND HEALTH ISSUES

It is extremely important to consider the safety and health of everyone who may be involved in, or come into contact with, waste being handled, treated or discharged. The hazards can be physical, chemical and biological. In many cases, there may be a wide range of wastes to consider. Those which constitute a risk should be clearly differentiated from non-hazardous waste. Chapter 8 deals with the principal areas for concern.

REFERENCES IN CHAPTER 2

1. *Industrial Waste Management: A CEFIC Approach to the Issue*, 1989, (CEFIC, Brussels, Belgium), available from CIA.
2. CIA, 1989, *Responsible Care*, A programme designed to help improve the chemical industry's performance in the fields of health, safety, environment, etc.
3. DoE and Foreign & Commonwealth Office, 1994, *Sustainable Development — The UK Strategy* (HMSO, UK) series number — Cm2426.
4. *CEFIC Guidelines on Environmental Reporting for the European Chemical Industry*, 1993 (CEFIC, Brussels. Belgium), available from CIA.
5. Jones, D.G., 1995, *Environmental Standards Certification Kit* (Gee Publishing Ltd, London, UK).
6. Lennon, T., 1995, Environmental insurance, *Environmental Protection Bulletin*, 38: 29–32.
7. Kolluru, R.V., 1995, Minimising EHS risks and improving the bottom line, *Chem Eng Prog*, 91 (6): 44–52.
8. Moilanen, T. and Martin, C., 1995, *Financial Evaluation of Environmental Investments* (IChemE, Rugby, UK).
9. *Guidelines for Safe Warehousing*, 1993 (CIA, UK).
10. *Storage of Packaged Dangerous Substances*, 1986, Guidance note CS17 (HSE, UK).

3. WASTE LAW AND REGULATORY AUTHORITIES

3.1 INTRODUCTION

There is a complex legal framework governing the production, storage, treatment, disposal and transport of waste. It is vitally important that any company operating in the process industries is familiar with this framework and with the regulatory authorities who have an enforcement role. The regulatory authorities are not there only to keep the law but also to offer advice and assistance. This chapter reviews the law pertinent to waste management and disposal (see Table 3.1, page 22). A good general reference work on environmental law has been written by Ball and Bell[1]. Safety and health law is covered in Section 8.2, page 176.

In the past, legal controls have been applied separately to solid, liquid and gaseous environmental media. However, the recognition that waste substances may be transferred between these media has led to the development of the concept of best practicable environmental option (BPEO). In the United Kingdom, the embodiment of this concept in the regime of Integrated Pollution Control (IPC) introduced by Environmental Protection Act 1990 means that it is no longer possible in legal terms to view waste in any form in isolation. To deal with the law pertaining to waste disposal and waste management comprehensively, it is therefore necessary to cover much of the field of environmental law.

The UK Department of the Environment (DoE) and Welsh Office produced a document in December 1995 entitled *Making Waste Work: A Strategy for Sustainable Waste Management in England and Wales* principally concerned with non-radioactive solid waste[2]. Under the Environment Agency, it will lead to a new statutory waste management policy which will replace the waste disposal plans drawn up by the Waste Regulation Authorities.

The principal legal control over waste in the UK is exercised through a number of statutes, such as the Control of Pollution Act 1974, EPA 1990, and the Water Resources Act 1991, and the detailed statutory instruments — that is, regulations — issued under these acts. However, the so-called common law (or case law) is also important; this is a body of law arising out of decisions made in the courts which are often binding on subsequent courts considering the same or similar factual situations.

TABLE 3.1
Overview of Chapter 3

3.2, page 24 Definitions of waste	Special Waste Regulations 1996, the Controlled Waste Regulations 1992 and Directive Waste 91/156 which will unify the EU definition
3.3, page 31 Land use planning	Town and Country Planning Act 1990 affects planning permissions and the need for environmental assessments on the use of land connected with waste treatment and disposal
3.4, page 31 Pollution control and waste regulation	The Environmental Protection Act 1990 is the principal environmental statute in the UK based on Integrated Pollution Control and best available techniques not entailing excess cost. It is implemented in a series of regulations including: • the Environmental Protection (Duty of Care) Regulations 1991; • the Enviromental Protection (Prescribed Processes and Substances) Regulations 1991; • the Enviromental Protection (Duty of Care) Regulations 1991; • the Waste Management Licensing Regulations 1994.
3.5, page 44 Carriage of waste	The Control of Pollution (Amendment) Act 1989 covers registration of waste carriers and the Controlled Waste (Registration of Carriers and Seizure of Vehicles) Regulations 1991 and Disposal of Controlled Waste (Exceptions) Regulations 1991 are relevant
3.6, page 45 Discharges to sewer	The Water Industry Act 1991 controls discharges to sewer and the Trade Effluents (Prescribed Processes and Substances) Regulations 1989 (as amended)
3.7, page 46 Discharges to controlled waters	The Water Resources Act 1991 deals with discharges to controlled waters
3.8, page 46 Radioactive substances	The Radioactive Substances Act 1993 covers the accumulation and disposal of radioactive waste
3.9, page 47 Common law	Adds to legislation in civil actions through case law the areas of nuisance, trespass, fraud and contracts

Continued opposite

TABLE 3.1 (continued)
Overview of Chapter 3

3.10, page 48 Regulatory authorities	The Environment Agency combines the following former regulatory authorities into a single body: • NRA (dealt with water environment); • HMIP (dealt with compliance and infringement, enforcement notices, prohibition notices and prosecution); • WDAs.
3.11, page 48 EC legislation	As a member of the EU, the UK is subject to EC legislation on the definition of waste, transfrontier shipments, landfills and incineration. There are still differences in detail between the EU states.
3.12, page 57 US Environmental law	Federal Regulations have produced a comprehensive multimedia approach

The European Community (EC) is also a source of relevant legislation, some of which is directly applicable in the United Kingdom, while other EC legislation is implemented by UK statutes. The UK can also be a party to international law. There are differences in the law and the regulatory authorities in England and Wales, Scotland and Northern Ireland. In general, the major Acts apply throughout the UK. However, the regulations and the Regulatory Authorities may differ, particularly in Scotland. Scottish Law relevant to the environment is outlined in Brodies[3]. Where there are significant differences these are highlighted in the text.

There are various guides referenced in this chapter which are more suitable for the informed lay person and aim to ensure that everyone is aware of, and understands, their legal responsibilities. Croner[4] has recently introduced a publication detailing important case law, which is updated regularly. Whilst it is important to achieve a degree of familiarity with waste law, it should be emphasized that guides such as these do not serve as a substitute for professional legal advice. There are a number of legal practices which specialize in environmental law and some of these can be identified through the UK Environmental Law Association (see Appendix 1 for contact details).

Another important aspect of the legal process is the 'expert witness'. Clearly, the selection of the right individual to support a company's position can be critical to the outcome of a case. There are a number of eminent people who have become established as expert witnesses from academia and consultancies

— see, for example, the *UK Register of Expert Witnesses*[5]. The legal practice itself may have a preferred expert, or the IChemE can provide a list of consultants[6].

The traditional 'command and control' approach of using solely legislation and enforcement for environmental issues is now under major review. The additional use of economic instruments such as charges and taxation is seen as a powerful tool for encouraging producers and consumers to be more environmentally responsible and to raise revenue for specific projects. The introduction of a landfill levy (see Section 7.5, page 162) is an example of an economic instrument designed to correct perceived market distortions and promote alternative waste management techniques.

3.2 DEFINITIONS OF WASTE

Many descriptions of waste — such as poisonous, toxic and so on — have no basis in law. There are few legal definitions of waste. The most important are described in this section.

3.2.1 WASTE

The statutory definition of waste is given in Section 75 of EPA 1990, whereby waste:
- 'includes (a) any substance which constitutes a scrap material or an effluent or other unwanted surplus substance arising from the application of any process and (b) any substance or article which requires to be disposed of as being broken, worn out, contaminated or otherwise spoiled';
- presumes that '... anything which is discarded or otherwise dealt with as if it were waste shall be presumed to be waste unless the contrary is proved'.

This definition does not, however, reflect the definition of waste in EC legislation. With the goal of introducing a common definition of waste for the Community, the Council of the European Communities issued Directive 91/156 which amends the existing Directive 74/492 ('the Framework Directive') that defined waste. Thus the term 'waste' has been significantly redefined under European Community law. The Waste Management Licensing Regulations 1994 amend the definition in Section 75 of EPA 1990 so that all references to 'waste' in the existing statutes now include a reference to 'Directive waste', as from 1 May 1994.

3.2.2 EC DIRECTIVE WASTE

The 1991 EC Directive 91/156 defines waste as any object or substance set out in Annex 1 (replicated in the UK 1994 Waste Management Licensing Regulations),

TABLE 3.2
Substances or objects which become waste when discarded

1	Production or consumption residues not otherwise specified in this table
2	Off-specification products
3	Products whose date for appropriate use has expired
4	Materials spilled, lost or having undergone other mishap, including any materials, equipment etc., contaminated as a result of the mishap
5	Materials contaminated or soiled as a result of planned actions — for example, residues from cleaning operations, packing materials, containers
6	Unusable parts — for example, reject batteries, exhausted catalysts
7	Substances which no longer perform satisfactorily — for example, contaminated acids, contaminated solvents, exhausted tempering salts
8	Residues of industrial processes — for example, slags, still bottoms
9	Residues from pollution abatement processes — for example, scrubber sludges, baghouse dusts, spent filters
10	Machining or finishing residues — for example, lathe turnings, mill scales
11	Residues from raw materials extraction and processing — for example, mining residues, oil field slops
12	Adulterated materials — for example, oils contaminated with PCBs
13	Any materials, substances or products whose use has been banned by law
14	Products for which the holder has no further use — for example, agricultural, household, office, commercial and shop discards
15	Contaminated materials, substances or products resulting from remedial action with respect to land
16	Any materials, substances or products which are not contained in the above categories

shown here as Table 3.2. The waste list is illustrative only and not exhaustive. The DoE Guidance in Circular 11/94 therefore states that in order for a substance or object to be waste it must fall into one of the categories set out in this list and:
- be discarded, disposed of or got rid of by the holder;
- be intended to be discarded, disposed of or got rid of by the holder;
- be required to be discarded, disposed of or got rid of by the holder.

The DoE Circular gives guidance on what is meant by 'discarded', 'disposed of' and 'intended to discard'.

3.2.3 CONTROLLED WASTE

'Waste' is generally dealt with under UK legislation as 'controlled waste' which means 'household, industrial and commercial waste or any such waste' (Section 75(4) of EPA 1990). It can only be disposed of via licensed disposal sites or with the necessary consents.

Household waste arises from a domestic property, caravan, residential home, educational establishment, hospital or nursing home, and does not fall within the scope of this book.

It is worth noting that the volume of clinical waste has grown rapidly in recent years and chemical engineers are now involved in its management, treatment and disposal.

Commercial waste is generated from premises (not factories) used for a trade or business or for sport, recreation or entertainment. It excludes household, industrial, mine and quarry and agricultural waste. It is not regarded generally as process industry waste and is not considered in this book. However, it should be borne in mind that a company's office premises, which may be remote from the manufacturing factory, will generate commercial waste which must be managed.

3.2.4 INDUSTRIAL WASTE

Industrial waste originates from a factory (defined under the UK Factories Act 1961) or premises which provide the public with:
- transport services;
- gas, electricity, water or sewerage services;
- postal or telecommunications services.

Schedule 3 of the Controlled Waste Regulations 1992 defines the types of waste which are to be treated as 'industrial waste', as listed in Table 3.3.

3.2.5 SPECIAL WASTE

Largely as a result of the EC Hazardous Waste Directive 91/689/EC (see Section 3.11.3, page 51), on 1 September 1996 the Special Waste Regulations 1996 came into force revoking and replacing the Control of Pollution (Special Waste) Regulations 1980. The Regulations cover the keeping, treatment or disposal of controlled wastes which are dangerous or difficult to manage. Detailed guidance on the Regulations is given in the Department of Environment Circular 6/96[7].

TABLE 3.3
Waste to be treated as industrial waste

1	Waste from premises for maintaining vehicles, vessels or aircraft, not being waste from a private garage to which Paragraph 4 of Schedule 1 applies
2	Waste from a laboratory
3	• Waste from a workshop or similiar premises not being a factory within the meaning of Section 175 of the Factories Act 1961 because the people working there are not employees or because the work there is not carried on by way of trade or for purposes of gain • In this paragraph 'workshop' does not include premises at which the principal activities are computer operations or the copying of documents by photographic or lithographic means
4	Waste from premises occupied by a scientific research association approved by the Secretary of State under Section 508 of the Income and Corporation Taxes Act 1988
5	Waste from degrading operations
6	Waste arising from tunnelling or from any other excavation
7	Sewage not falling within a description in Regulation 7 which: • is treated, kept or disposed of in or on land, other than by means of a privy, cesspool or septic tank • is treated, kept or disposed of by means of mobile plant, or • has been removed from a privy or cesspool
8	Clinical waste other than: • clinical waste from a domestic property, caravan, residential home or from a moored vessel used wholly for the purpose of living accommodation • waste collected under Section 22(3) of the Control of Pollution Act 1974, or • waste collected under Section 89, 92(9) or 93
9	Waste arising from any aircraft, vehicle or vessel which is not occupied for domestic purposes
10	Waste which has previously formed part of any aircraft, vehicle or vessel and which is not household waste

Continued overleaf

TABLE 3.3 (continued)
Waste to be treated as industrial waste

11	Waste removed from land on which it has previously been deposited and any soil with which such waste has been in contact, other than: • waste collected under Section 22(3) of the Control of Pollution Act 1974 • waste collected under Sections 89, 92(9) or 93
12	Leachate from a deposit of waste
13	Poisonous or noxious waste arising from any of the following processes undertaken on premises used for the purposes of a trade or business: • mixing or selling paints • sign writing • laundering or dry cleaning • developing photographic film or making photographic prints • selling petrol, diesel fuel, paraffin, kerosene, heating oil or similar substances • selling pesticides, herbicides or fungicides
14	Waste from premises used for the purposes of breeding, boarding, stabling or exhibiting animals
15	Waste oil, waste solvent or (subject to Regulation 7(2)) scrap metal other than: • waste from a domestic property, caravan or residential home • waste falling within Paragraphs 3 to 6 of Schedule 1 In this paragraph: • 'waste oil' means mineral or synthetic oil which is contaminated, spoiled or otherwise unfit for its original purpose, and • 'waste solvent' means solvent which is contaminated, spoiled or otherwise unfit for its original purpose
16	Waste arising from the discharge by the Secretary of State of his duty under Section 89(2)
17	Waste imported into the UK
18	Tank washings or garbage landed in the UK In this paragraph: • 'tank washings' has the same meaning as in Regulation 2 of the Control of Pollution (Landed Ship's Waste) Regulations 1987, and • 'garbage' has the same meaning as in Regulation 1(2) of the Merchant Shipping (Reception Facilities for Garbage) Regulations 1988

TABLE 3.4
Waste substances to be treated as 'special waste'

02	Agricultural, horticultural, hunting, fishing and aquaculture primary production, food preparation and processing
03	Wood processing and the production of paper, cardboard, pulp, panels and furniture
04	Leather and textile industry
05	Petroleum refining, natural gas purification and pyrolytic treatment of coal
06	Inroganic chemical processes
07	Organic chemical processes
08	Manufacture, formulation, supply and use of coatings (paints, varnishes and vitreous enamels), adhesives, sealants and printing inks
09	Photographic industry
10	Power station and other combustion plants
11	Inorganic waste with metals from metal treatment and the coating of metals; non-ferrous hydro-metallurgy
12	Shaping and surface treatment of metals and plastics
13	Oil wastes (except edible oils)
14	Organic substances employed as solvent
15	(blank)
16	Waste not otherwise specified, i.e. miscellaneous
17	Construction and demolition
18	Human or animal health care and/or related research
19	Waste treatment facilities, off-site waste water treatment plants and the water industry
20	Municipal wastes and similiar commercial, industrial and institutional wastes

Note — the above are the major classes of special waste and the full waste code has six digits of which the above are the first two digits. A more comprehensive listing can be found in *Croner's Waste Management Guide*[9] and in the Special Waste Regulations 1996 (Part 1, Schedule 2).

The definition of special waste is now based on the list of hazardous waste from EC Directive 94/904/EC but also includes certain other substances. Table 3.4 highlights the main categories of special waste. If a waste is listed, and in addition, has any of the following properties it is classed as Special Waste:

- explosive — under the effect of flame or is more sensitive to shock or friction than dinitrobenzene;
- oxidizing — highly exothermic reactions when in contact with other substances, particularly flammable substances;
- highly flammable:

- liquids with flashpoint below 21°C;
- flammable gases;
- solid waste which may readily catch fire after brief contact with a source of ignition and which continue to burn or to be consumed after removal of the source of ignition;
- waste which may become hot and catch fire after brief contact with air at ambient temperatures and without any application of energy;
- wastes which, in contact with water or damp air, evolve highly flammable gases in dangerous quantities;
- flammable — liquids with flashpoint between 21 and 55°C;
- irritant — non-corrosive wastes which, through immediate, prolonged or repeated contact with the skin or mucous membrane, can cause inflammation;
- harmful — wastes which may involve limited health risks;
- toxic — wastes which may involve serious, acute or chronic health risks or even death;
- carcinogenic — wastes which may cause cancer or increase its incidence;
- corrosive — wastes which may destroy living tissue on contact;
- infectious — wastes containing micro-organisms or their toxins which cause disease in man or other living organisms;
- teratogenic — wastes which may induce non-hereditary birth defects or increase their incidence;
- mutagenic — wastes which may induce hereditary genetic defects or increase their incidence;
- toxic gas release — wastes which release toxic gases in contact with water, air or acid;
- ecotoxic — wastes which present immediate or delayed risks for the environment.

Wastes which after disposal produce a leachate or other substance with any of the above characteristics will be classed as special. Characteristics of wastes should be determined as for chemical products according to the Chemicals (Hazard Information and Packaging for Supply) Regulations 1994 (CHIPS 2). Further details on CHIP 2 are available from HSE[8].

Finally, if a waste does not appear in Table 3.4 but possesses any of the following properties, it will be also defined as special:
- liquid with flashpoint less than 21°C;
- irritant, harmful, toxic, carcinogenic or corrosive;
- prescription-only medicine.

3.3 LAND USE PLANNING

In general, any use of land (apart from agriculture or defence use) requires planning permission (or an exemption) under the Town and Country Planning Act 1990. Applications for planning permission must be made to the district council and advertised locally. In appropriate cases, which fall under the Town and Country Planning (Environmental Assessment) Regulations 1988, the application must be accompanied by an environmental statement which fully describes the proposed development and may include measures to be taken to avoid or reduce any adverse environmental impacts. The submitted statement will be the outcome of an environmental impact assessment of the proposed development.

The facilities under which assessment is mandatory are listed in the Town and Country Planning (Assessment of Environmental Effects) Regulations 1988 (as amended). An environmental impact assessment is required for all developments for the disposal of special waste, including incinerators, chemical waste treatment and landfill facilities.

Other proposals may require an environmental assessment when the proposed development is one which would, in the opinion of the relevant planning authority, be likely to have 'significant' effects on the environment by virtue of matters such as nature, size or location. Should an application be made without provision of an environmental statement, there is a risk that the application would not be accepted and time periods for its consideration by the relevant determining body will not begin to run. Assistance as regards matters that need to be considered are set out in a guide provided by the DoE entitled *Environmental Assessment — A Guide to the Procedures* (known as the 'Blue Book').

3.4 POLLUTION CONTROL AND WASTE REGULATION

EPA 1990 put pollution control in Britain on a radically new statutory basis. Part I of the Act establishes two regimes: Integrated Pollution Control (IPC) for the most potentially polluting or technologically complex industrial processes; and Local Authority Air Pollution Control (LAAPC) for others. Part II deals with waste on land and, among other things, introduces a 'duty of care' on persons concerned with controlled waste and a new system of waste management licensing. Part III consolidates the law regarding statutory nuisances.

There is also provision under Section 157 of EPA 1990 for directors, managers and other similar officers to be liable to prosecution. Charges can be brought against such individuals where it can be shown that any offence committed by the company was done with the 'consent or connivance of, or to have been attributable to any neglect on the part of' those individuals.

The Environmental Protection (Prescribed Processes and Substances) Regulations 1991 as amended specify the industrial processes which are 'prescribed' by the Secretary of State as requiring an Authorization under Part I of EPA 1990. The regulations distinguish between the industrial processes which are subject to local authority control for air pollution only (the 'Part B' processes) and those with the potential to make significant discharges into more than one environmental medium (the 'Part A' processes), which require an IPC authorization from the Environment Agency (formerly HMIP).

EPA 1990 requires operators or proposed operators of an IPC or LAAPC process to obtain an authorization from the appropriate enforcing authority in order to continue operating or to begin operations. The enforcing authority is the Environment Agency (formerly HMIP) for IPC and the district and borough councils and port health authorities for LAAPC.

These regulations are also the means by which the Secretary of State prescribes specific substances, which are those most potentially harmful when released into the three media and are subject to special requirements to ensure that the best available techniques not entailing excessive cost (BATNEEC) are used to prevent their release or to minimize releases. The substances prescribed for the three media are shown in Table 3.5, page 33. Note that these substances are only those deserving extra consideration: all substances which might cause harm if released into any medium are additionally subject to the use of BATNEEC to render them harmless.

3.4.1 INTEGRATED POLLUTION CONTROL

IPC considers discharges from industrial processes to all environmental media (namely air, land and water) as a whole. The idea stems from the concept of best practicable environmental option (BPEO), originated by the Royal Commission on Environmental Pollution. BPEO is the pollution control strategy which best considers the overall impact on air, land and water. This includes transfers between media, such as the use of a scrubber to remove pollutants from the air thereby giving rise to a liquid effluent stream. EPA 1990 attempts to embody the concept of BPEO by requiring that industrial processes with the potential to make significant discharges into more than one environmental medium are covered by the IPC regime, which requires that the operators of the process demonstrate that their process meets BPEO.

When making an application for an IPC authorization, applicants must present an assessment to show that their process represents the BPEO — that is, releases from the process will be controlled so as to have the least impact on the environment as a whole. Though this does not mean a detailed environmental impact assessment for all substances in each process option, it does mean an

TABLE 3.5
Release into the air, water and land

Air	• oxides of sulphur and other sulphur compounds • oxides of nitrogen and other nitrogen compounds • oxides of carbon • organic compounds and partial oxidation products • metals, metalloids and their compounds • asbestos (suspended particulate matter and fibres), glass fibres and mineral fibres • halogens and their compounds • phosphorus and its compounds • particulate matter
Water	• mercury and its compounds • cadmium and its compounds • all isomers of hexachlorocyclohexane • all isomers of DDT • pentachlorophenol and its compounds • hexachlorobenzene • hexachlorobutadiene • aldrin • dieldrin • endrin • polychlorinated biphenyls • dichlorvos • 1,2-dichloroethane • all isomers of trichlorobenzene • atrazine • simazine • tributyltin compounds • triphenyltin compounds • trifluralin • fenitrothion • azinphos-methyl • malathion • endosulfan
Land	• organic solvents • azides • halogens and their covalent compounds • metal carbonyls

Continued overleaf

TABLE 3.5 (continued)
Release into the air, water and land

Land (continued)	• organometallic compounds • oxidizing agents • polychlorinated dibenzofuran and any congener thereof • polychlorinated dibenzo-*p*-dioxin and any congener thereof • polyhalogenated biphenyls, terphenyls and naphthalenes • phosphorus • pesticides* • alkali metals and their oxides and alkaline earth metals and their oxides

*Any chemical substance or preparation prepared or used for destroying any pest including those used for protecting plants or wood or other plant products from harmful organisms; regulating the growth of plants, giving protection against harmful creatures; rendering such creatures harmless; controlling organisms with harmful or unwanted effects on water systems, buildings or other structures, or on manufactured products; or protecting animals against ectoparasites

assessment of the main areas that the process might impact on (globally, regionally or locally) with a justification of the technique proposed to be used in that context.

3.4.2 LOCAL AUTHORITY AIR POLLUTION CONTROL

As noted already, the Environmental Protection (Prescribed Processes and Substances) Regulations 1991 also set out the industrial processes which are subject to local authority control for air pollution only (the 'Part B' processes). For these processes, authorization is sought through the environmental health department of the local authority in which the process is located. A number of guidance notes on some 80 processes have been produced by the Department of the Environment including general advice on authorizations, applications, registers and appeals (see Appendix 6). The notes are reviewed every four years.

In addition to the authorizations, local authorities must deal with the 'statutory nuisance' provisions of the EPA 1990 (see Section 3.4.7, page 43). These cover air pollution and noise problems. The local authorities undertake a significant amount of air quality monitoring, particularly in smoke control areas.

There is a memorandum of understanding between the local authorities and the Environment Agency which aims to ensure speedy two-way flow of information on any problems with a process or breaches of an authorization.

3.4.3 BEST AVAILABLE TECHNIQUES NOT ENTAILING EXCESSIVE COST (BATNEEC)

Authorizations issued under Part I of EPA 1990 oblige the operator of the process to use the best available techniques not entailing excessive cost (BATNEEC):

- for preventing the release of substances prescribed for any environmental medium into that medium or, where that is not practicable by such means, for reducing the release of such substances to a minimum and for rendering harmless any such substances which are so released; and
- for rendering harmless any substances which might cause harm if released into any environmental medium.

The obligation to use BATNEEC may be embodied in specific conditions in an authorization and may cover:

- the mode of operation of the plant;
- the pollution abatement technology to be used; and
- the explicit limits on the amount and concentration of substances which may be discharged.

BATNEEC may encompass a wide range of factors such as management methods, personnel, design, layout and maintenance. Inevitably, there will be aspects of the process which are too detailed or obscure to be covered explicitly by the authorization conditions and in such cases there remains a residual duty on the process operator to prevent or minimize the release of prescribed substances by the use of BATNEEC.

In addition to specific conditions relating to BATNEEC, an authorization may contain conditions specified in directives from the Secretary of State and other conditions as appear to the enforcing authority to be appropriate.

Once issued, the authorization has the force of law through enforcement notices, etc. Criminal prosecutions may be brought by the enforcing authority against an operator for failure to obtain, or comply with the conditions in, an authorization. This is a major departure from the previous system under which an operator had only a general duty to use 'best practicable means' to avoid pollution. A person is guilty of an offence for not complying with the conditions in an authorization or of enforcement/prohibition notices. Offences on summary conviction carry a maximum fine of £20,000. In the Crown Court the fine is unlimited and/or up to two years' imprisonment.

The meaning of BATNEEC is explained in Government policy guidance as:

- *best* means the *most effective* means of preventing or minimizing or rendering harmless pollution emissions. There may be more than one set of techniques that achieve comparable effectiveness — that is, there may be more than one set

MANAGEMENT OF PROCESS INDUSTRY WASTE

of best techniques.
- *available* means procurable by the operator of the process. This does not necessarily mean that the process is in general use but it does require general accessibility. Just because it is available outside the United Kingdom, for example, does not mean that it is unavailable. Nor is there implied a competitive supply market.
- *techniques* refers not only to the technology of the process but also how it is operated. The two components of 'techniques' may be broken down into 'staff' and 'buildings'. As to staff, the number, qualifications, training and supervision must be considered, and as to buildings, their design, construction, layout and maintenance.
- *not entailing excessive cost* is perhaps a more difficult concept and there is doubt about its precise legal meaning. With *new* processes it is presumed that the best available techniques (BAT) will be used. However, where the cost of doing so will be excessive, given the nature of the industry and the environmental protection to be achieved, then considerations of whether or not the cost is excessive become relevant.

As regards the upgrading of existing processes, guidance notes indicate the timescales over which it is generally appropriate to upgrade to new plant standards. All authorizations are required to be revised at least once every four years.

3.4.4 GUIDANCE NOTES

The Department of the Environment has produced general guidance notes for local authorities on the operation of LAAPC and a large number of guidance notes on specific processes. Process operators should be able to assess from these notes what constitutes BATNEEC for different categories of process.

Guidance notes are available and issued to inspectors setting out, for each IPC process:
- why the process is subject to IPC and what prescribed substances might be released;
- any standards relevant to the process — for example, environmental quality standards or international requirements such as for sulphur dioxide and nitrogen oxides under the EC Large Combustion Plants Directive;
- the standards which the Environment Agency believes can be achieved by pollution control in that process;
- a technique or range of techniques suitable for achieving those standards;
- what would be considered as a substantial change and require an alteration to the authorization.

Although the guidance notes are intended to ensure that there is consistency in what is considered BATNEEC, the onus is still on the operator to demonstrate that the process uses BATNEEC.

3.4.5 THE DUTY OF CARE

The most important changes brought about by Part II of EPA 1990 as far as the producer of waste is concerned is the introduction of a duty of care. It is applicable to controlled waste controlled by:
- importers and producers;
- carriers, keepers and treaters;
- disposers (including recyclers);
- brokers.

In other words, it applies to anyone who has control of waste from the time of its creation to the time of its final disposal as enshrined in the popular phrase 'cradle to grave'. The 'duty of care' applies to all 'controlled' waste whether solid or liquid.

Such persons are required to take all 'such measures applicable to (them) in that capacity as are reasonable in the circumstances':
- to prevent any person from contravening provisions of EPA 1990 which prohibit the unauthorized or harmful deposit, treatment or disposal of waste;
- to prevent the escape of waste from the control of the authorized person, or that of any other person;
- on the transfer of the waste, to make sure:
— that the transfer is only to an authorized person or to a person for authorized transport purposes; and
— that there is transferred a written description of the waste which will enable other persons to avoid a contravention of the provisions of the EPA 1990 cited above and to comply with the duty of care with respect to the escape of waste.

The term 'authorized persons' is defined in Section 34(3) of the EPA 1990 and includes:
- Waste Collection Authorities — usually District Councils in England and Wales. In Scotland, they are the Islands or District Councils;
- holders of waste management licences under EPA 1990;
- persons registered as carriers of controlled waste under Section 2 of the Control of Pollution (Amendment) Act 1989 and the Controlled Waste (Registration of Carriers and Seizure of Vehicles) Regulations 1991.

'Authorized transport purposes' encompasses the transportation of waste within the same premises and for certain purposes connected with the importation and/or export of waste.

EPA 1990 is essentially a framework under which regulations and guidance notes can be made to ensure that amendments may be made quickly to meet changing needs and technology. Therefore, Section 34, which provides for the 'duty of care', does not go into detail. This was left to the Environmental Protection (Duty of Care) Regulations 1991. These regulations impose requirements on any person who is subject to the 'duty of care', in making or retaining documents and furnishing of copies of them. Breach of these regulations is a criminal offence.

Under the regulations, the transferor and the transferee are required to complete a transfer note which goes with a written description of the waste being transferred. The transfer note should contain such information as:
- type and quantity of waste being transferred;
- type of storage being used;
- time and place of transfer;
- name and address of the transferor and the transferee;
- transferor as either the producer or importer of the waste;
- category of the transferor and the transferee;
- authorized transport purpose (if any and which applies);
- certain additional information.

The transfer note and the written description of the waste must be kept by both the transferor and the transferee for a minimum of two years from the date of the transfer. Any documents kept under these regulations must be produced to the Environment Agency (formerly Waste Regulation Authority) on request. Further details on the system of consignment notes is given in Appendix 5, page 191.

The code of practice associated with the regulations is aimed at setting out practical guidance for waste holders subject to the 'duty of care'[10]. It sets out a series of steps which would usually be sufficient to meet the 'duty of care'. The code of practice, although not having legislative effect, is admissible in civil and criminal proceedings. The legal obligation, however, is to comply with the 'duty of care' rather than with the code. Table 3.6, page 39, shows the Codes summary check-list. Figure 3.1, page 41, is the Controlled Waste Transfer Note from the Code.

3.4.6   WASTE MANAGEMENT LICENSING

The keeping, treatment and disposal of waste are licensed activities under Part II of EPA 1990 which relates to waste on land. A new waste licensing system was brought into force on 1 May 1994 and operates within the following framework:
- the Waste Management Licensing Regulations 1994 and Department of the

TABLE 3.6
Summary check-list

This table draws together in one place a simple check-list of the main steps that are normally necessary to meet the duty of care. As with the code as a whole, this does not mean that completing the steps listed here is all that needs to be done under the duty of care. The check-list cross-refers to key sections of the code for fuller advice.

		Refer to paragraphs
(a)	Is what you have waste? If yes,	Introduction and the Appendix
(b)	Is it controlled waste? If yes,	1.2
(c)	While you have it, protect and store it properly	2.1–2.4
(d)	Write a proper description of the waste, covering:	1.7–1.9
	• any problems it poses	1.3–1.6 and 1.10
	and, *as necessary to others who might deal with it later,* one or more of:	
	• the type of premises the waste comes from	1.11–1.12
	• what the waste is called	1.13
	• the process that produced the waste	1.14–1.15
	• a full analysis	1.16–1.17
(e)	Select someone else to take the waste; they must be an authorized person or authorized for transport purposes. As such they should be one or more of the following and must prove that they are:	
	• a registered waste carrier	3.6–3.8
	• exempt from waste carrier registration	3.9–3.12
	• a waste manager licensed to accept the waste	3.13–3.16
	• exempt from waste management licensing	3.17–3.18
	• a waste collection authority	3.5
	• in Scotland, a waste disposal authority operating under the terms of a resolution	3.20–3.21
	• authorized transport purposes	3.3
(f)	Anyone arranging on behalf of another person for the disposal or recovery of controlled waste must be	2.5–2.7
	• a registered waste broker	3.4, 3.19. 4.10
	• an exempt waste broker	3.4
(g)	Pack the waste safely when transferring it and keep it in your possession until its is to be transferred	2.5–2.7

Continued overleaf

MANAGEMENT OF PROCESS INDUSTRY WASTE

TABLE 3.6 (continued)
Summary check-list

(h)	Check the next person's credentials when transferring waste to them	3.17–3.12 3.15–3.19
(i)	Complete and sign a transfer note	Annex C, C.3–C.4
(j)	Hand over the description and complete a transfer note when trasnferring the waste	1.7–1.9 Annex C, C.3–C.4
(k)	Keep a copy of the transfer note signed by the person the waste was given to, and a copy of the description, for two years	Annex C, C.5
(l)	When *receiving* waste, check that the person who hands it over is one of those listed in (e), or the producer of the waste, obtain a description from them, complete a transfer note and keep the documents for two years	4.2–4.10 5.4
(m)	Whether transferring *or* receiving waste, be alert for any evidence or suspicion that the waste handled is being dealt with illegally at any stage, in case of doubt question the person involved and if not satisfied, alert the EA or the Scottish Protection Agency	5.5–5.11
(n)	Supplementary guidance for holders of scrap metal	Section 7

Environment Circular 11/94 (Welsh Office 26/94, Scottish Office 10/94);
- Waste Management Paper No 4;
- a charging scheme.

The aims of the Waste Management Licensing Regulations 1994 are to prescribe the:
- cases where a waste management licence is not required;
- form and content of applications in respect of waste management licences;
- offences that are relevant for the purposes of Section 74(3)(a) (see below) and the qualifications and experience required of a person for the purposes of Section 74(3)(b) (Fit and Proper Persons, also explained below);
- arrangements for appeals against the refusal of a licence or licence conditions;
- particulars that are to be entered into registers to be kept by waste regulation and collection authorities;
- descriptions of plant that are to be treated as being, or as not being, mobile plant;
- conditions which are, or are not, to be included in a licence.

Duty of Care: Controlled Waste Transfer Note

Section A - Description of Waste

1. Please describe the waste being transferred:

2. How is the waste contained?

    Loose ☐   Sacks ☐   Skip ☐   Drum ☐   Other ☐ → *please describe:*

3. What is the quantity of waste (number of sacks, weight etc):

Section B - Current holder of the waste

1. Full Name (BLOCK CAPITALS):

2. Name and address of Company:

3. Which of the following are you? (Please ✓ one or more boxes)

producer of the waste ☐	holder of waste disposal or waste management licence ☐ →	License number: Issued by:
importer of the waste ☐	exempt from requirement to have a waste disposal or waste management licence ☐ →	Give reason:
waste collection authority ☐	registered waste carrier ☐ →	Registration number: Issued by:
waste disposal authority (Scotland only) ☐	exempt from requirement to register ☐ →	Give reason:

Section C - Person collecting the waste

1. Full Name (BLOCK CAPITALS):

2. Name and address of Company:

3. Which of the following are you? (Please ✓ one or more boxes)

	authorized for transport purposes ☐ →	Specify which of those purposes:
Waste Collection Authority ☐	holder of waste disposal or waste management licence ☐ →	License number: Issued by:
Waste Disposal Authority (Scotland only) ☐	exempt from requirement to have a waste disposal or management licence ☐ →	Give reason:
	registered waste carrier ☐ →	Registration number: Issued by:
	exempt from requirement to register ☐ →	Give reason:

Section D

1. Address of place of transfer/collection point:

2. Date of transfer:

3. Time(s) of transfer (for multiple consignments, give 'between' dates):

4. Name and address of broker who arranged this waste transfer (if applicable):

5. Signed: Signed:

    Full name: Full name:
    (BLOCK CAPITALS)   (BLOCK CAPITALS)

    Representing: Representing:

Figure 3.1   Duty of care: controlled waste transfer.

EPA 1990 sets out the criteria to be taken into account by the Environment Agency when considering whether or not to grant a licence. If planning permission relating to the proposed activity is in force, the Environment Agency may reject an application for a waste management licence if it considers that the activity may give rise to pollution of the environment or harm to human health. Furthermore, where the Environment Agency is not satisfied that the applicant is a 'fit and proper person' then it must reject the application.

An applicant may be treated as not being 'fit and proper' if:
- the management of the licence will not be in the hands of a technically competent person;
- the applicant does not have the finance or has not made the necessary financial provision to discharge the licence obligations;
- the applicant or relevant person has committed a relevant offence.

A person is considered technically competent if, and only if, they hold the relevant Certificate of Technical Competence (COTC) from the Waste Management Industry Training and Advisory Board (WAMITAB). Details are given in Table 1 of the Regulations.

There is further detail in Waste Management Paper No 4 as to the Department of the Environment's interpretation of the requirement in Section 74(3)(b) of EPA 1990 that a site be 'in the hands of a technically competent person'.

A person may not be 'fit and proper' if convicted of a relevant offence. A conviction will not necessarily render a person no longer fit and proper — the Environment Agency has a discretion, guidance on which is set out in Waste Management Paper No 4. Relevant factors include the gravity and number of offences. Relevant offences for this purpose are contained within Regulation 6. The relevant offences include a wide range of waste pollution and nuisance offences.

Where a company, or a director or other officer of that company, is convicted of a relevant offence then that conviction may serve to invalidate a related company sharing the same director or officer from holding a waste management licence. Conversely, the conviction of a company may disqualify a director of that company from holding a licence. Where one or more companies will be holding waste disposal licences, careful consideration needs to be given to the organization of those companies to minimize the potential for other companies and individuals being penalized in the event of a successful prosecution.

The potential categories of licensable activities in EPA 1990 are very wide when interpreted literally and so the regulations give specific exemptions to activities which the Government does not intend to be encompassed by the licensing regime. The exemptions cover:

- activities subject to other legal controls — for example, effluents are the subject of discharge consents under the Water Resources Act 1991, discharges subject to the Integrated Pollution Control system (except waste sent for final disposal to land);
- other activities which are excluded from licensing but are still subject to Section 33(l)(c) of EPA 1990 which precludes the treating, keeping or disposing of controlled waste in a manner likely to cause pollution of the environment or harm to humans.

EPA 1990 provides that a licence can only be surrendered if the Environment Agency accepts that the condition of the land is unlikely to cause pollution of the environment or harm to human health. It is anticipated, therefore, that waste management licences cannot be surrendered until some 30 years after operations have ceased on site.

Proposals for a registration scheme for waste brokers were issued by the Department of the Environment on 10 August 1994. Up to 250 businesses are affected, including some whose activities in arranging for waste imports into the UK have caused concern in recent years.

3.4.7 STATUTORY NUISANCE

Part III of EPA 1990 contains provisions on 'nuisance', which is broadly defined as an interference with someone's (or the public's) enjoyment or comfort and which could be prejudicial to health. These may be pertinent to problems encountered in the disposal of waste. There are three types of nuisance: public and private nuisance which are part of common law and statutory nuisance — that is, where a particular nuisance has been made so by statute. Part III of EPA 1990 (Sections 79–85) aimed to consolidate existing definitions of statutory nuisances from other statutes — for example, The Public Health Act 1936 —into a single list, although some others still exist. Statutory nuisances are defined to include:

- premises in such a state as to be prejudicial to health or a nuisance;
- smoke emitted from premises so as to be prejudicial to health or a nuisance;
- fumes or gases emitted from premises so as to be prejudicial to health or a nuisance;
- any dust, steam, smell or other effluvia arising on industrial, trade or business premises and being prejudicial to health or a nuisance;
- any accumulation or deposit which is prejudicial to health or a nuisance.

A local authority is under a duty to inspect its area from time to time to detect any statutory nuisance, and where a complaint is made of a statutory nuisance by a person living within the local authority's area, the authority is under a duty to investigate the complaint. Where the local authority is satisfied

that a statutory nuisance exists, or is likely to occur or recur, the authority is required to serve an abatement notice. The notice may:
- require the abatement of the nuisance or prohibit or restrict its occurrence or reoccurrence;
- require the execution of works and the taking of other steps, as may be necessary for any of those purposes.

This is an important power allowing a local authority to act in anticipation of a nuisance. There is an important additional defence in cases where nuisance arises on industrial, trade or business premises. In such cases, it is a defence to prove that the best practicable means were used to prevent or counteract the effects of the nuisance. 'Practicable' takes into account the current state of technical knowledge and financial implications for the operator. The 'means' to be employed include the design, installation, maintenance and manner and periods of operation of plant machinery, and the design, construction and maintenance of buildings and structures.

Failure to comply with an abatement notice empowers the local authority to abate the nuisance and do whatever may be necessary in execution of the notice and recover any costs reasonably incurred in so doing from the person by whose act or default the nuisance was caused. It is also an offence not to comply with an abatement notice, without reasonable excuse. Where the authority is of the opinion that proceedings for an offence would afford an inadequate remedy, it may proceed directly to the High Court to obtain an injunction.

Another feature is that private individuals aggrieved by a statutory nuisance may apply to a court for an abatement order. The order which a Magistrates Court is empowered to make in response to a complaint lodged by an individual is in the same terms as that for an abatement notice by a local authority.

3.5   CARRIAGE OF WASTE

The Control of Pollution (Amendment) Act 1989 (CoPA) provides the legal framework for registration of waste carriers. The Controlled Waste (Registration of Carriers and Seizure of Vehicles) Regulations 1991 were made under this act. They make it an offence to transport controlled waste in the course of business or for profit unless registered with the Environment Agency. There are a few exemptions in Regulation 2.

The Environment Agency may refuse registration if the applicant has been convicted of any of the offences in Schedule 1 of the regulations which mainly relate to pollution control legislation. The Environment Agency is also given powers to apply for the seizure of vehicles which have been used for fly-tipping of waste.

3.6 DISCHARGES TO SEWER

The UK Water Industry Act 1991 (WIA) is the statute which controls discharges of industrial liquid waste to sewer. Under Section 118 of the Water Industry Act 1991, an application must be made to the statutory undertaker (SU) for a consent to discharge any trade effluent from premises into the public sewer system. In England and Wales, the SU is the relevant one of the ten major water public limited companies. The procedure for obtaining a consent is set out in Chapter III to Part IV of the Act.

On first approaching the SU, a company may be required to give preliminary details — for example, the nature of the waste and industry, volume and rate of flow. The SU will inform the company of likely conditions and give a summary of the charging scheme and controls. The company then submits a trade effluent notice for consent by the undertaker.

The SU may specify in the consent, the sewer into which the effluent is to be discharged, maximum daily quantities, highest rate of discharge and the nature and composition of that discharge. Red List substances, as listed in Schedule 1 to the Trade Effluents (Prescribed Processes and Substances) Regulations 1989 (as amended), may be prohibited from discharge above background concentration by the SU. Appeals are made to the Director of the Office of Water Services (OFWAT). If acceptable to the SU then the undertaker refers the acceptable limits to the Secretary of State (Environment Agency) for determination. This must be done within two months of the trade effluent notice. The conditions required by the Secretary of State to be included in the Consent or Agreement must be attached by the SU.

Agreements to discharge, as an alternative to consents, can still be entered into under S129(1) WIA. These should however be regarded as commercial contracts and are becoming less common.

For processes which are prescribed for Integrated Pollution Control an application is made to the Environment Agency. Statutory Consultees are informed, including the SU. A separate Consent or Agreement will probably be needed from the SU in addition to the IPC authorization.

Charges may be imposed as a condition of the consent or as a term of the agreement under which the discharge is made.

In Scotland, where an occupier or prospective occupier of trade premises proposes to make a new discharge of trade effluent from those premises into the sewerage system of a local authority, the consent of that local authority must be obtained (Section 26 of the Sewerage (Scotland) Act 1968). On granting consent the authority may impose similar conditions to those under the WIA.

3.7 DISCHARGES TO CONTROLLED WATERS

The Water Resources Act 1991 (WRA) deals with the functions and responsibilities of the former National Rivers Authority (NRA) — now the Environment Agency — and replaces corresponding sections of the Water Act 1989. It relates particularly to discharges to controlled waters.

Under Section 85 of the WRA, a person is guilty of an offence by causing or knowingly permitting any poisonous, noxious or polluting matter to be discharged:
- into controlled waters;
- through a pipe into the sea outside the seaward limits of controlled waters;
- in contravention of a prohibition from a building or other fixed plant onto or into any land;
- into any waters of a lake or pond which are not inland waters.

Under Section 88, no offence is committed if a discharge to controlled waters is made in accordance with an appropriate licence, including an IPC authorization (see Section 3.4.1, page 32). When an application to operate a prescribed process is made to the Environment Agency, statutory consultees are informed and conditions incorporated if deemed appropriate. The conditions for discharge are included in the authorization. For discharges to controlled waters, the conditions imposed are not more relaxed than those required by the Environment Agency.

All discharges to controlled waters (including territorial waters, coastal waters, inland waters and ground water) which are not prescribed processes require an application to be made to the Environment Agency.

In Scotland, under the Control of Pollution Act 1974, control of pollution in rivers and coastal waters was the responsibility of the River Purification Boards (RPBs) — now the Scottish Environment Protection Agency (SEPA). They may give consent to discharges of poisonous, noxious or polluting matter, or sewage or trade effluent into controlled waters (defined as above) subject to conditions. Applications for IPC in Scotland are made to SEPA.

The Environment Agency has the power to carry out works to prevent polluting matter from entering controlled waters, and to recover the costs from whoever caused or knowingly permitted the polluting matter to be there (under Section 161).

3.8 RADIOACTIVE SUBSTANCES

Although outside the scope of this book, it should be noted that under the provisions of the UK Radioactive Substances Act 1993 an authorization is required for the accumulation of radioactive waste with a view to its subsequent disposal (Section 7.14 of the act) and for the disposal of any radioactive waste (Section 7.13 of the act). Applications for authorizations should be made to the Environment Agency.

3.9 COMMON LAW

Under English law, industrial operators may find themselves open to civil actions arising from the effects of their emissions even when they have not breached statutory controls. The common law provides a number of possible actions that may be brought against persons causing or permitting pollution to occur and, in some cases, against the owners of land from which pollution emanates even if they did not specifically cause it themselves.

The relevant aspects of common law under English and Welsh law are:

THE RULE IN RYLANDS V FLETCHER

This Rule applies in the case of a 'non-natural' use of land and escape from that land causing loss to an injured person or neighbouring land. The person liable is the person who caused the problem (not necessarily the current owner) and liability is 'strict' — that is, there is no need for the plaintiff to show fault on the part of the defendant but must show reasonable foreseeability. The historic activities of a company can therefore come back to haunt it.

NUISANCE

An action will lie against a person who, as owner or occupier of land, causes interference with the comfort and convenience (enjoyment) of neighbouring land — for example, through fire, leachate, gas, noise, smells. The person in occupation at the time the nuisance was committed could be liable even after selling on. The new owner who knows of and adopts the nuisance could be liable.

TRESPASS

Trespass is the unjustifiable intrusion onto the land of another. It involves direct physical intrusion — for example, deposit of waste. The person who commits the trespass remains liable.

FRAUD

The deliberate concealment of an environmental problem — for example, contamination — will leave a vendor liable in fraud (deceit) even after the vendor has sold the land or company.

NEGLIGENCE

A person may be liable in negligence due to physical injury or loss resulting from that person's activities, including leaving land in a dangerous condition. A previous owner will also be liable for economic loss where there is a breach of a term of a contractual relationship between the previous owner and the injured party to leave the site properly filled and restored. The liability remains with the person who commits the negligent act or omission.

CONTRACT

- There may be a breach of an obligation under the contract to deliver land or a business in a specified condition — for example, operating lawfully.
- There may be a specific obligation to indemnify the other party for losses or costs resulting from the condition of a site handed over under a waste management licence or lease or arising out of the sale of polluted land.
- There may be liability in misrepresentation where a selling party innocently or fraudulently makes an incorrect statement which induced the other party to enter into the contract.

3.10   REGULATORY AUTHORITIES

The responsibility for environmental regulation in England and Wales rests with the Secretary of State for Environment in the Department of the Environment (see Table 3.7, page 49).

The Environment Agency operates on a regional basis for all its functions. However, there are separate and overlapping boundaries for internal handling of pollution control and water management. Customers have a single point of local contact. The key Director posts include Environmental Strategy, Operations, Personnel, Pollution Prevention and Control, Water Management and Finance. Of these the Director of Pollution Prevention and Control is most relevant to waste management as these responsibilities cover pollution prevention and control for water, waste, air and land on an integrated basis.

3.11   EC LEGISLATION

As a member state of the European Union (EU), the UK is subject to European Community (EC) environmental law[11]. The EU's environmental policy originates in the so-called Action Programmes. Since 1973, five Action Programmes on the environment have been adopted, leading to some 300 pieces of legislation. The Fifth Programme's objective is sustainable development and runs

TABLE 3.7
Regulatory authorities

Department of the Environment	Oversees all aspects of environmental regulation. It also provides policy and advice on environmental issues.
The Environment Agency	Deals with major industrial operations to ensure compliance of IPC including BATNEEC and BPEO and liaises with local authorities environmental health departments
	Most frequent infringements due to: • lack of (inadequate) operational procedures to prevent or minimize emission; often peripheral to main process — for example, bulk liquid storage and handling, vehicle loading and unloading, effluent handling and treatment • delays in completing improvement programmes by specified date • emissions to air • failure of maintenance procedures
	Responsible for controlling water pollution and resources — works under the Water Resources Act 1991 and some 20 EC Directives. Its main activity is setting up and implementing Catchment Management Plans.
	Minor industrial operations and discharges to air and noise
Waste Collection Authorities	Collects household and commercial waste, and industrial waste if approved by WDA
Waste Disposal Authorities	Dispose of certain controlled wastes at its own sites and regulate disposal of all controlled wastes (covers site licensing and monitoring of special waste transport and disposal). Note — in Scotland, the WDA's can have a much broader range of activities, including operation of recycling plants, incinerators and energy from waste plants in and out of their areas.

Continued overleaf

TABLE 3.7 (continued)
Regulatory authorities

Local Authority Waste Disposal Companies	The EPA 1990 separated disposal from local authority control with existing WDA's required either to set up LAWDCs (but with important restrictions) or use private waste management companies.
Health and Safety Commission and Executive	Protection of the health and safety of workers and the public — it is specifically responsible for major accident hazards, pesticides, genetically modified organisms and hazardous chemicals
Water plcs (sewage undertakers)	Discharges to sewers under the Water Industry Act 1991
The Scottish Environment Protection Agency	Similiar to the Environment Agency but there are differences and exemptions in the regulatory framework.

from 1993 to 2000. It aims to introduce a broad range of economic and fiscal incentives/disincentives and civil liabilities to correct perceived market distortions.

A significant resolution on waste management was adopted by EU environmental ministers in March 1990. It established a hierarchy with waste minimization at the top — that is, most preferable — followed by recycling and reuse, with the optimization of waste treatment and disposal at the bottom. This approach is increasingly reflected in EC legislation and UK policy.

3.11.1   EC DIRECTIVES, REGULATIONS AND DECISIONS

EC law is enacted through a series of primary treaties (such as the Treaty of Rome) and secondary legislation in the form of regulations, directives and decisions.

Regulations are of general application and binding on all member states. They are directly applicable in the sense that they create legal rights and duties enforceable in the member states and therefore no national legislation is necessary to implement them.

Directives are not of general application. They are addressed to member states and can be directed to one or all of several member states. They are binding as to the results to be achieved, but the method of implementation is left to the individual state. EU legislation regarding the environment usually takes the form of directives.

Decisions are binding in their entirety upon those to whom they are addressed, whether private individuals or member states. They have been used in the environmental field in connection with international conventions and with certain procedural matters. The most important regulations and directives for hazardous waste are discussed below.

3.11.2 WASTE FRAMEWORK DIRECTIVE

The 1975 'Framework' Directive (75/442) established general rules and principles to ensure that treatment and disposal of waste in all EU member states is undertaken without endangering human health or the environment. The designation of competent authorities was required, with provision for these authorities to prepare waste disposal plans. Permits were required for undertakings which treat, store or tip waste on behalf of third persons. A supervisory system was required for undertakings which dispose of their own waste. This directive was substantially amended by Directive 91/156.

3.11.3 HAZARDOUS WASTE DIRECTIVE

The Hazardous Waste Directive (91/689) is a general directive which will be followed by so-called daughter directives. Also, like the Framework Directive referred to in Section 3.11.2, the directive affects the definition of waste. Broad categories of substances which are deemed to be hazardous are set out in an annex and this is supplemented by a detailed list drawn up by an expert committee (in 94/904).

The directive contains provisions relating to the safe disposal of hazardous waste, its description, packaging and record-keeping by competent authorities, producers, transporters and disposers and provisions on the mixing of hazardous waste with other waste. Authorities are required to draw up plans for the management of hazardous waste: these may be separate from, or incorporated into, more general plans.

Although implementation of the directive has been subject to delay due to the need to agree the detailed list, the directive has now repealed the Directive on Toxic and Dangerous Waste (78/319) as of 27 June 1995, which described a system of competent authorities, permitting, inspecting and record keeping, by reference to the 'polluter pays' principle. All of these aspects are covered in the new directive. Implementation of the 91/689 has resulted in new regulations on special waste in the United Kingdom (see Section 3.2.5, page 26).

3.11.4 TRANSFRONTIER SHIPMENT OF WASTE

EC Regulation 259/93 on the supervision and control of waste within, into or out of the EU came into operation on 6 May 1994. The regulation replaces and extends existing EC legislation on transfrontier shipments of waste within, into or out of the EU. Its purpose is, among other things, to give effect to the Basel

Convention on the control of transboundary movements of hazardous waste and its disposal. However, the regulation applies to all waste as defined in other EC legislation. The main provisions of the new regulation lay down procedures and controls for waste movements involving different groups of countries, with separate provisions being made for wastes intended for disposal and those intended for recovery.

The regulation contains procedures for the notification of waste movements, and acknowledgement/objection procedures for regulatory authorities to follow. The authority in the country receiving the waste must also authorize the shipment. Conditions may be set and further information requested by the authorities. The waste must also be checked at the end of its journey.

In the UK, the regulation has been implemented by the Transfrontier Shipment of Waste Regulations 1994. Further details are given in Section 7.8 of this book, on page 173.

3.11.5 PROPOSED DIRECTIVE ON THE LANDFILLING OF WASTE

The European Council adopted a common position on the landfilling of waste in October 1995. The proposal is markedly different from the first proposal in 1991. It will lay down strict rules on the type of waste that can be accepted at various site categories and sets down a procedure for the classification of sites. Uniform procedures will have to be adopted and pollution monitoring will become compulsory. Certain wastes will not be acceptable in a landfill. These include:
- combustible or flammable waste;
- explosive wastes;
- hospital and other infectious clinical wastes;
- other wastes which do not fulfil the criteria set down in the directive.

The proposal is designed to limit the disposal of waste at landfills and place emphasis on waste reduction and reuse or recycling.

3.11.6 DIRECTIVE ON INCINERATION OF HAZARDOUS WASTE

The main European Council Directive on the Incineration of Hazardous Waste 94/67/EC adopts an integrated approach towards environmental protection covering not only air pollution but also the protection of the soil, surface water and groundwater. To achieve its aim of minimizing effects on the environment and on human health, the proposed directive sets out operating conditions and emission limit values for hazardous waste incineration plants within the European Community.

Hazardous wastes are those defined in the Directive on Hazardous Waste (91/689), incineration of which may give rise to harmful emissions including organic compounds, hydrogen chloride, hydrogen fluoride, heavy metals, dioxins and furans. The directive does not say that such wastes must be incinerated nor does it restrict its scope to specialized incinerators burning only hazardous waste.

The operation of incinerators is to require a permit which must be granted only if 'the incineration plant is designed, equipped and will be operated in such a manner that the appropriate preventive measures against environmental pollution will be taken'.

The directive contains requirements covering the delivery of wastes to the plant to ensure that the composition and characteristics of the waste are known to the incinerator operator. There should be a description of the waste, it must be weighed, samples taken and all documents verified.

The directive sets out detailed mandatory design and operational features — for example, that the temperature in the combustion chamber must be raised to 850°C and kept at that for at least two seconds in the presence of 6% oxygen (if waste with more than 1% halogenated organic substances is incinerated the temperature must be 1100°C). Emission limit values for the plant's gaseous discharges are specified in Article 7 of the directive. The incineration of hazardous waste as regular or additional fuel for any industrial process is subject to the same limit values.

In order to protect the aquatic environment, the discharge of waste waters from exhaust gas cleaning is to be 'limited as far as possible'. Additionally, inline with the goal of rational use of energy, the heat generated by the incineration of hazardous waste should be used as far as possible instead of wasted.

Member states are to bring into force the laws required to comply with the directive before 31 December 1996. Existing hazardous waste incinerators will then have up to three years and six months to meet the requirements.

3.11.7   DIRECTIVE ON PACKAGING AND PACKAGING WASTE
This directive was adopted in December 1994 and includes all packaging waste from industrial, commercial and other sources. It sets out targets for recovery — that is, recycle, compost or incineration with energy recovery. There are provisions in the UK Environment Act 1995 which require reuse, recycling and recovery of packaging materials. This will be effected by an industry scheme.

3.11.8   GREEN PAPER ON CIVIL LIABILITY FOR WASTE
The EC's initial draft Directive on Civil Liability for Damage Caused by Waste published in September 1989 produced sharply conflicting responses and did not progress beyond the draft stage. The draft directive aimed to impose 'strict

TABLE 3.8
Hazardous waste management systems[†]

	Austria	Denmark	France	Germany
Status				
Date of main legislation	1983	1972/92	1975	1972/86/90
Registration/licensing[a]				
Collector/transporters	L	[b]	R	L
Treatment/disposal contractors	L	L	R	L
Transportation				
Manifest system	Yes	Yes	New	Yes
Control over import	Yes	Yes	Yes	Yes
Control over export	No	Yes	Soon	Yes
Permitting				
Storage	Yes[b]	Yes	Yes	Yes
Treatment	Yes	Yes	Yes	Yes
Disposal	Yes	Yes	Yes	Yes
Have all facilities been permitted?	No	Yes	Yes	Yes
Planning and siting				
Is there a national strategy/plan?	Yes	Yes	No	No
Are authorities required to produce a plan?	Yes	Yes	No	Yes
Has this been done?	Yes	Yes	No	Yes
Abandoned sites				
Is there a national inventory?	New	Yes	Yes	Yes
Is there a clean-up program?	No	Yes	[e]	[e]

[a] L — licensing scheme, implying investigation by authorities
R — registration, implying simply being listed in a register
[b] Mainly under the Trade Act, not under the Hazardous Waste Act
[c] Provincial or state responsibility

[†] Based on material from LaGrega, M., 1994, *Hazardous Waste Management*, by permission of The McGraw-Hill Companies

WASTE LAW AND REGULATORY AUTHORITIES

	Italy	Netherlands	Spain	Sweden	UK	USA
	1982/84	1979	1986/88	1975	1972/74	1976
	L	No	L	L	No	R
	L	L	L	L	No	R
	Yes	Yes	Yes	Soon	Yes	Yes
	Yes	Yes	Yes	Yes	Soon	Yes
	Yes	Yes	Yes	Yes	Soon	Yes
	Yes	Yes	Yes	Yes	Soon	Yes
	Yes	Yes	Yes	Yes	Yes	yes
	Yes	Yes	Yes	Yes	Yes	Yes
	Yes	Yes	No	Yes	Yes	No
	Yes	Yes	Yes	No	No	No
	Yes	Yes	Yes	No	Yes	c
	Yes	Yes	No	No	Partial	Partial
	Yes[d]	Yes	No	Yes	No	Yes
	Yes[d]	Yes	No	Yes	No	Yes

[d] Partial

e Although there is no formal, nationwide cleanup program, the cleanup of individual sites is proceeding

TABLE 3.9
Control over treatment or disposal options[†]

	Austria	Denmark	France	Germany
Direction of waste				
To particular site(s)	No	Yes	No	No
To particular option(s)	[a]	Yes	No	Yes
Powers exist, in reserve	No	—	Yes	No
Prohibition of certain options for particular wastes				
National regulations	[a]	Yes	Yes	Yes
Control via site permits				
Strong national standards mean effective prohibition for certain waste	Yes/No[b]	Yes	Yes	Yes/No[b]

[a] Recommendations are made, but these are not mandatory
[b] Strong controls exist in principle, but in practice there are wide local variations in what is/is not permitted at individual sites
[c] System not yet in place

[†] Based on material from LaGrega, M., 1994, *Hazardous Waste Management*, by permission of The McGraw-Hill Companies

liability' on the deemed producer of waste for up to 30 years from the date on which the incident giving rise to the damage occurred. Under 'strict liability' it is not necessary to show negligence or fault by the waste producer. This approach is favoured by those who believe it increases the incentives for greater care in environmental activities.

A subsequent Green Paper in March 1993 was issued to stimulate further discussion on accountability and compensation for damage caused to the environment. It covers both 'strict liability' and 'fault-based liability'. The major problem with the latter is the difficulty in proving any wrongful action(s). In balance the EC still favours 'strict liability'. It also favours the establishment of joint compensation mechanisms (sector based) where liability cannot be attributed to a single party. However, it is unclear where the discussions on the Green Paper will lead.

Italy	Netherlands	Spain	Sweden	UK	USA
No	No	No	Yes	No	No
No	No	No	Yes	No	Yes
No	No	Yes	—	Yes	No
Yes	Yes	Yes	Yes	No	Yes
Yes	Yes	Yes/No[c]	Yes	No	Yes

3.11.9 OTHER EU COUNTRIES

The legislative position in other EU countries is summarized in Tables 3.8 (pages 54–55), 3.9 and 3.10 (pages 58—59).

3.12 US ENVIRONMENTAL LAW

It is only possible in this book to outline the US system briefly. It is valuable, however, to look at the US multimedia approach as it has been the spur for many international environmental legislative systems. The US Environmental Protection Agency administers the federal regulations enabled by the US Congress. There are also state and local laws which are relevant but these are not considered in this brief review. Table 3.11, page 60, summarizes the main federal regulations.

TABLE 3.10
Definitions of hazardous waste[†]

	Austria	Denmark	France	Germany
Is there a legal definition?	Yes	Yes	Yes	Yes
Purpose of definition				
• control over transport	Yes	Yes	Yes	Yes
• control over treatment/disposal	Yes	Yes	Yes	Yes
Type of definition				
• list of waste	Yes	Yes	No	Yes
• list of substances	No	Yes	Yes	No
• list of processes	No	No	Yes	No
• concentrations	No	No	No	No
Criteria:	Yes	Yes	Yes	Yes
• toxicity of waste	Yes	Yes	Yes	Yes
• toxicity of extract	No	Yes	Yes	No
• ignitability/flammability	No	Yes	Yes	No
• corrosiveness	No	Yes	Yes	No
• reactivity	No	Yes	Yes	No
Special rules:				
• mixing rule	Yes	Yes	Yes	Yes
• residue rule	No	Yes	Yes	No
Exclusions:				
• small generators[a]	No	No	100 kg	No
• wastewater	Yes	No	Yes	Yes
• sewage sludge	Yes	No	No	No
• mining waste	Yes	No	Yes	Yes
• agricultural waste	Yes	No	Yes	Yes

[a] Quantity is that per month below which a producer is exempt from the regulations
[b] Partial exclusion for wastewater treated exclusively in permitted treatment tanks. Wastewater treated in surface impoundments or lagoons is controlled as a hazardous waste.
[c] Mining waste excluded pending further study

[†] Based on material from LaGrega, M., 1994, *Hazardous Waste Management*, by permission of The McGraw-Hill Companies

WASTE LAW AND REGULATORY AUTHORITIES

Italy	Netherlands	Spain	Sweden	UK	USA
Yes	Yes	Yes	Yes	Yes	Yes
Yes	Yes	No	Yes	Yes	Yes
Yes	Yes	Yes	Yes	No	Yes
No	Yes	Yes	Yes	No	Yes
Yes	Yes	Yes	Yes	Yes	Yes
Yes	Yes	Yes	Yes	No	Yes
Yes	Yes	No	No	No	Yes
Yes	No	Yes	Yes	Yes	Yes
Yes	—	Yes	Yes	No	Yes
No	—	Yes	No	Yes	Yes
No	—	Yes	No	Yes	Yes
No	—	Yes	No	Yes	Yes
No	—	Yes	No	No	Yes
Yes	Yes	No	No	No	Yes
No	No	No	No	No	Yes
No	No	No	No	No	100 kg
Yes	Yes	Yes	No	Yes	[b]
Yes	No	Yes	Yes	Yes	No
No	No	Yes	No	Yes	Yes[c]
No	No	Yes	No	Yes	No

TABLE 3.11
US environmental regulations

Regulation	Explanation
Resource Conservation and Recovery Act (RCRA) 1976	RCRA defined hazardous waste and set up Office of Solid Waste in US EPA to regulate 'cradle to grave' control in solid and hazardous waste generation, treatment, storage (including underground), transport and disposal
Hazardous and Solid Waste Amendments 1984	Amended RCRA, the register covers land disposal prohibitions, minimum technology requirements, permitting, corrective action program, inspection and enforcement and underground storage tanks
Federal Register of Regulations 1980	Produced under RCRA cover lists of hazardous materials, notification procedures for waste producers, regulations on generation, transport, treatment, storage and disposal
Comprehensive Environmental Response, Compensation and Liability Act (CERCLA) 1980	CERCLA set up $1.6 billion fund for identification, analysis and clean-up of inactive hazardous waste treatment, storage and disposal sites
Superfund Amendment and Reauthorization Act 1986	CERCLA rewritten and $8.5 billion fund set up for clean-up of old disposal sites — note the influence of 1984 Bhopal tragedy in 'Right–to–Know' provisions, list of Superfund sites established and remedial actions planned. Set up Toxic Release Inventory for raw environmental data.
Toxic Substances Control Act (TSCA) 1976	TSCA permits EPA to gather data and regulate testing, manufacture, distribution, import and export of chemical substances, including PCBs and asbestos. Prior to manufacture and use, all chemicals must be listed by EPA under TSCA.
Occupational Safety and Health Act 1970	Principal regulation dealing with safety and health at work including exposure to materials such as wastes.

Continued opposite

TABLE 3.11 (continued)
US environmental regulations

Regulation	Explanation
Clean Water Act (CWA) 1972, 1977, 1987	CWA covers physical, chemical and biological integrity of US waters. Also known as Public Law 92–500, it is responsible for the National Pollutant Discharge Elimination System aimed at zero discharges.
Safe Drinking Water Act 1974, 1986	Covers the safety of drinking water supplies by establishing and enforcing national drinking water standards, including organics, viral and bacterial contamination.
Clean Air Act 1970, 1977, 1990	Covers prevention and control of air pollution and its effect on public health and welfare. It established the concept of National Ambient Air Quality Standards for regulated pollutants, New Source Performance Standards for specific processes and National Emission Standards for Hazardous Air Pollutants. Limits imposed by Prevention of Significant Deterioration.

REFERENCES IN CHAPTER 3
1. Ball, S. and Bell, S., 1995, *Environmental Law*, 3rd edition (Blackstone Press Ltd, UK).
2. *Making Waste Work: A Strategy for Sustainable Waste Management in England and Wales*, 1995 (DoE/Welsh Office, UK).
3. Brodies, 1992, *Scots Law and the Environment* (Brodies Environmental Law Group, T&T Clark, Edinburgh, UK).
4. *Croner's Case Law*, 1995 (Croner Publications Ltd, UK).
5. *UK Register of Expert Witnesses*, 1995, 8th edition (JS Publications, Newmarket, Suffolk, UK).
6. IChemE, 1996, *List of Consultants 1996/97* (IChemE, Rugby, UK).
7. *Environmental Protection Act 1990: Part II, Special Waste Regulations 1996*, Circular 6/96, 1996 (DoE, UK).
8. *CHIP 2 for Everyone*, 1995, HS(G) 126 (HSE, UK).
9. *Croner's Waste Management Guide*, 1991 (Croner Publications Ltd, UK).
10. DoE/Scottish office, 1991, Environmental Protection Act 1990: Waste Management, the Duty of Care; a code of practice (HMSO, UK).
11. Handler, T. (ed), 1994, *Regulating the European Environment* (Chancery Law Publishing, London, UK).

4. PROCESS AND WASTE CHARACTERIZATION

4.1 INTRODUCTION

The characterization of wastes can be divided between analytical procedures and routine monitoring. The selection of the best practicable environmental option (BPEO) for dealing with waste requires a thorough classification of the nature of all wastes generated by a process. Particular attention must be paid to the critical components of the waste stream — for example, heavy metal concentrations in wash waters, total chemical oxygen demand in liquid effluents and soluble components in any solid waste. These components define the treatment/disposal route and the maximum disposal rate.

Routine monitoring is aimed at ensuring compliance with discharge consents and for control purposes. In general, monitoring systems are less accurate than analytical procedures and only give a partial characterization of the waste.

In the UK, the requirements to characterize waste are set out in the Environmental Protection Act 1990 (EPA 1990), the EC Directive 91/156, Special Waste Regulations 1996, Controlled Waste Regulations 1992 and the Waste Management Licensing Regulations 1994. However, there is little direct guidance on the characterization methods to be used. Table 4.1 gives an overview of Chapter 4, page 64.

4.2 PROCESS CHARACTERIZATION

Understanding the nature of the process(es) under consideration goes a long way towards defining the likely waste issues. Useful general descriptions of a wide range of processes can be found in Kirk-Othmer's *Encyclopedia of Chemical Technology*[1] and Shreve's book[2] on the chemical process industries.

A major distinction can be made between inorganic and organic operations. Care must be taken, however, as notional inorganic operations can still generate significant amounts of organic wastes from cleaning/maintenance solvents and materials, and from packaging materials. In general, there are more options in dealing with organic wastes than inorganic wastes, as discussed in Chapter 6.

The first question is whether a process is 'prescribed'. The Environmental Protection (Prescribed Processes and Substances) Regulations 1991 as

TABLE 4.1
Overview of Chapter 4

4.2, page 63 Process characterization	• process description • new process considerations • 'prescribed' processes • 'prescribed' substances
4.3, page 65 Waste assessment / survey	• process flow diagram (PFD) • mass balance • quantify all waste streams and critical waste component(s) in each stream • sampling • variations in quantity • composition • include peripheral areas such as handling raw materials and products • storage systems • drainage systems
4.4, page 78 Characterization / analysis	• key parameters • analytical techniques
4.5, page 78 Monitoring	• off-line and continuous • standard test methods
4.6, page 83 Database	• data specifications and structures
4.7, page 84 Marking and labelling	• regulations and UN guidelines

amended specify the industrial processes which are 'prescribed' by the Secretary of State as requiring an authorization under Part I of EPA Act. The regulations distinguish between the industrial processes which are subject to local authority control for air pollution only ('Part B' processes) and those with the potential to make significant discharges into more than one environmental medium ('Part A' processes) that require an IPC authorization from the Environment Agency.

A timetable for implementation is included in the regulations. For any new plant or a substantially changed existing one to operate a prescribed IPC process, an authorization has been needed since 1 April 1991 and existing processes fell under IPC only gradually according to the timetable. Applications for all processes must have been submitted by 31 January 1996. All LAAPC processes are now subject to the authorization requirement.

The next question is whether any of the materials handled in the process are 'prescribed'. These regulations are also the means by which the Secretary of State prescribes specific substances, which are those most potentially harmful when released into the three environmental media and are subject to special requirements to ensure that BATNEEC is used to prevent or minimize releases. The substances prescribed for the three media are shown in Table 3.5, page 33. Note that these substances are only those deserving extra consideration: all substances which might cause harm if released into any medium are additionally subject to the use of BATNEEC to render them harmless.

Croner's Waste Management Guide[3] has a comprehensive section of 'waste charts' which set out for each waste material/compound an overview of the physical, chemical, medical and environmental hazards, treatment and disposal options. There are also notes on handling precautions, labelling, storage and legislation.

4.3 WASTE ASSESSMENT/SURVEY

A waste assessment or survey must be undertaken to identify waste streams and current waste management practices. Figure 4.1, page 66, gives an overview of any industrial process with an emphasis on waste. The assessment/survey reveals and quantifies all of the items highlighted in Figure 4.1. Note that due attention is paid to the procurement of raw materials and dispatch of product, as a significant number of legislative infringements have occurred in these areas (see Section 3.10, page 48). All raw materials used in the process are included, such as water, cleaning fluids, maintenance materials and replacement items or equipment. To stimulate and aid those undertaking the waste assessment/survey, Tables 4.2 (page 67) and 4.3 (page 68) show typical ways in which waste occurs. It is important to be alert to all potential sources of waste and to question the validity of all information gathered.

The basis for the assessment is the process flow diagram (PFD). The survey aims to verify that the PFD mirrors the actual plant and, where there are differences, the PFD is updated and the reasons investigated. From the PFD, all waste streams can be listed and given a unique reference ID. Strictly, it is necessary to include all outputs from a process on the PFD. This may be difficult, however, for fugitive and evaporative losses.

Additional difficulties are created when some streams contain trace quantities. Note that the PFD simply shows the current 'end-of-pipe' waste streams which emanate from any existing treatment systems. It may be valuable

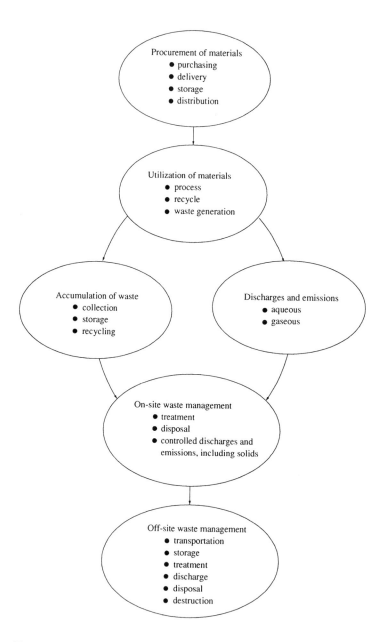

Figure 4.1   Tracking inputs, products and wastes. (Source: Wentz, C., 1989, *Hazardous Waste Management*, reproduced by permission of The McGraw-Hill Companies.)

TABLE 4.2
Typical wastes from plant operations (source: US Environmental Protection Agency Manual and Guide)

Plant function	Location/operation	Potential waste material
Material receiving	Loading docks, incoming pipelines, receiving areas	Packaging materials, off-spec materials, damaged containers, inadvertent spills, transfer hose emptying
Raw materials and product storage	Tanks, warehouses, drum storage yards, bins, storerooms	Tank bottoms, off-spec and excess materials, spill residues, leaking pumps valves, tanks and pipes, damaged containers; empty containers
Production	Melting, curing, baking, distilling, washing, coating, formulating, reaction, materials handling	Washwater, rinse water, solvents, still bottoms, off-spec products, catalysts, empty containers, sweepings, ductwork clean-out, additives, oil, filters, spill residue, excess materials, process solution dumps, leaking pipes, valves, hoses, tanks and process equipment
Support services	Laboratories	Reagents, off-spec chemicals, samples, empty sample and chemical containers
	Maintenance shops	Solvents, cleaning agents, degreasing sludges, sand-blasting waste, caustic, scrap metal, oils, greases
	Garages	Oils, filters, solvents, acids, caustics, cleaning bath sludges, batteries
	Powerhouses/boilers	Fly ash, slag, tube clean-out material, chemical additives, oil, empty containers, boiler blowdown
	Cooling towers	Chemical additives, empty containers, cooling tower blowdown, fan tube oils

TABLE 4.3
Causes and controlling factors in waste generation (source: US Environmental Protection Agency Manual and Guide)

Waste/ origin	Typical causes	Operational factors	Design factors
Chemical reaction	• incomplete conversion • by-product formation • catalyst deactivation (by poisoning or sintering)	• inadequate temperature control • inadequate mixing • poor feed flow control • poor feed purity control	• proper reactor design • proper catalyst selection • choice of process • choice of reaction conditions
Contact between aqueous and organic phases	• condensate from steam jet ejectors • presence of water as a reaction by-product • use of water for product rinse • equipment cleaning • Spill clean-up	• indiscriminate use of water for cleaning or washing	• vacuum pumps instead of steam jet ejectors • choice of process • use of reboilers instead of steam stripping
Process equipment cleaning	• presence of residual material • deposit formation • use of filter aids • use of chemical cleaners	• excessive use of hazardous cleaners • drainage prior to cleaning • production scheduling to reduce cleaning frequency • switch from batch to continuous operation	• provide wiper blades for reactor and tank inner surface • use equipment dedication to reduce cross-contamination • design equipment and piping to minimize hold-up
Heat exchanger cleaning	• presence of residual material (process side) or scale (cooling water side) • deposit formation • use of chemical cleaners	• inadequate cooling water treatment • excessive cooling water temperature	• design for lower film temperature and high turbulence • controls to prevent cooling water from overheating

Continued opposite

PROCESS AND WASTE CHARACTERIZATION

TABLE 4.3 (continued)
Causes and controlling factors in waste generation
(source: US Environmental Protection Agency Manual and Guide)

Waste/ origin	Typical causes	Operational factors	Design factors
Metal parts cleaning	• disposal of spent solvents, spent cleaning solution, or cleaning sludge	• indiscriminate use of solvent or water	• choice between cold dip tank or vapour degreasing • choice between solvent aqueous cleaning solution
Metal surface treating	• dragout • disposal of spent treating solution	• poor rack maintenance • excessive rinsing with water • fast removal of workpiece	• countercurrent rinsing • fog rinsing • dragout collection tanks or trays
Disposal of unusable raw materials or off-spec products	• obsolete raw materials • off-spec products caused by contamination, improper reactant controls, inadequate pre-cleaning of equipment or workplace, temperature or pressure excursions	• poor operator training or supervision • inadequate quality control • inadequate production planning and inventory control of feedstocks	• use of automation • maximize dedication of equipment to a single function
Clean-up of spills and leaks	• manual material transfer and handling • leaking pump seals • leaking flange gaskets	• inadequate maintenance • poor operator training • lack of attention by operator • excessive use of water in cleaning	• choice of gasketing materials • choice of seals • use of welded or seal-welded construction

Continued overleaf

TABLE 4.3 (continued)
Causes and controlling factors in waste generation
(source: US Environmental Protection Agency Manual and Guide)

Waste/ origin	Typical causes	Operational factors	Design factors
Paint application	• overspray • colour change • clean-up	• use of solvent-based rather than water-based paint • spray angle, rate and overlap • paint solids content	• automization method (air, pressure or centrifugal) • electrostatic application • automate painting to improve application
Paint removal	• replacing worn coating • removing defective coating	• inadequate quality control • use of solvent strippers	• use abrasive or cryogenic stripping • use less hazardous

to backtrack through the process to identify the real origins of waste so that the subsequent phases of waste minimization, on-site waste treatment and off-site treatment/disposal are evaluated in the most unconstrained way. For example, it may be much more sensible to replace a current end-of-pipe treatment system with a new system that facilitates recycle than to add on an extra treatment unit.

The next step is to quantify each waste stream. There are various sources of useful information on the process which provide the starting point for the stream assessment as listed in Table 4.4. The US Environmental Protection Agency (EPA) has produced a number of worksheets as part of its waste minimization activities which can be used in the waste assessment[4,5] (see Figures 4.2–4.6, pages 72–76). This assessment encompasses times of start-up, shut-down and high, low and no operation modes to ensure complete characterization. It is likely that process variations are most noticeable in the waste streams rather than the final product stream.

An attempt to close the mass balance on the process is made to ensure that all waste has been identified and quantified correctly. Even under the best circumstances, however, the accuracy with which the mass balance can be closed is still likely to be inadequate from a waste point of view. It is for this reason that a survey of the process in which each waste stream is characterized and analysed is essential.

TABLE 4.4
Information for waste minimization assessments

Environmental information	Waste manifests and disposal records
	Waste analyses, flows and concentrations
	Wastewater discharge records and analyses
	Air emission records and analyses
	Compliance requirements
	Air emission limits
	Discharge consents
	Site licence controls
	Environmental assessment reports
	BS environmental management system, preparatory review
	Environmental audit reports
	Environmental health office data
Design information	Process descriptions
	Process flow diagrams
	Design and actual material and energy balances for production and pollution control processes
	Operating manuals
	Equipment lists, specifications and data sheets
	Piping and instrumentation diagrams
	Plot and elevation plans
	Equipment layouts and work flow diagrams
Raw materials and production information	Raw materials, product and intermediate specifications
	Material safety and environmental data sheets
	COSHH assessments
	Product and raw material inventory records
	Operator data logs and day books
	Operating procedures
	Production schedules
Economic information	Treatment and disposal costs for all forms of waste
	Water and sewer charges
	Product, utility, energy and raw material costs
	Operating and maintenance costs
	Departmental cost accounting reports
	Storage and transport costs
Other information	Company environmental policy statements
	Standard procedures and organizational charts
	Planning consents and conditions
	Personnel policy

MANAGEMENT OF PROCESS INDUSTRY WASTE

Firm _____	**Waste Minimization Assessment**	Prepared by _____
Site _____		Checked by _____
Date _____	Project No _____	Sheet 1 of 4 Page__of__

| WORKSHEET **9a** | **INDIVIDUAL WASTE STREAM CHARACTERIZATION** | ✿EPA |

1. Waste stream name/ID: _____ Stream number _____

   Process unit/operation _____

2. **Waste characteristics** (attached additional sheets with composition data, as necessary)

   ☐ gas   ☐ liquid   ☐ solid   ☐ mixed phase

   Density, lb/cuft _____ High heating value, Btu/lb _____

   Viscosity/consistency _____

   pH _____ Flash point _____ % water _____

3. **Waste leaves process as:**

   ☐ air emission   ☐ waste water   ☐ solid waste   ☐ hazardous waste

4. **Occurrence**

   ☐ continuous

   ☐ discrete

   discharge triggered by   ☐ chemical analysis

   ☐ other (describe)

   Type: ☐ periodic _____ length of period: _____

   ☐ sporadic (irregular occurrence)

   ☐ non-recurrent

5. **Generation rate**

   Annual _____ lbs per year

   Maximum _____ lbs per _____

   Average _____ lbs per _____

   Frequency _____ batches per _____

   Batch size _____ average _____ range

Figure 4.2   Example worksheet.

PROCESS AND WASTE CHARACTERIZATION

Firm _____	**Waste Minimization Assessment**	Prepared by _____
Site _____	Proc. Unit/Oper. _____	Checked by _____
Date _____	Project No _____	Sheet 2 of 4 Page __ of __

| WORKSHEET **9b** | **INDIVIDUAL WASTE STREAM CHARACTERIZATION** | ⊕EPA |

(continued)

6. Waste origins/sources

    Fill out this worksheet to identify the origin of the waste. If the waste is a mixture of waste streams, fill out a sheet for each of the individual waste streams.

    Is the waste mixed with other wastes?   ☐ Yes   ☐ No

    Describe how the waste is generated.

    _____
    _____
    _____
    _____
    _____
    _____
    _____
    _____
    _____
    _____
    _____
    _____
    _____
    _____
    _____
    _____
    _____
    _____
    _____
    _____
    _____
    _____
    _____

    Example: Formation and removal of an undesirable compound, removal of an unconverted input material, depletion of a key component (eg drag-out), equipment cleaning waste, obsolete input material, spoiled batch and production run, spill or leak cleanup, evaporative loss, breathing or venting losses, etc

Figure 4.3   Example worksheet.

Firm _____	**Waste Minimization Assessment**	Prepared by _____
Site _____	Proc. Unit/Oper. _____	Checked by _____
Date _____	Project No _____	Sheet 3 of 4 Page __ of __

| WORKSHEET **9c** | **INDIVIDUAL WASTE STREAM CHARACTERIZATION** | ✦EPA |

(continued)

Waste stream _____

7. **Management method**

   Leaves site in
   ☐ bulk _____
   ☐ roll off bins _____
   ☐ 55 gal drums _____
   ☐ other (describe) _____

   Disposal frequency _____

   Applicable regulations [1] _____
   _____

   Regulatory classification [2] _____
   _____

   Managed   ☐ onsite   ☐ offsite
                  ☐ commercial TSDF   _____
                  ☐ own TSDF   _____
                  ☐ other (describe)   _____
                                       _____

   Recycling ☐ direct use/re-use _____
                  ☐ energy recovery   _____
                  ☐ redistilled   _____
                  ☐ other (describe)   _____

                  reclaimed material returned to site?
                  ☐ Yes   ☐ No   ☐ used by others
                  residue yield _____
                  residue disposal/repository _____
                  _____

Note[1] list federal, state and local regulations (eg RCRA, TSCA, etc)
Note[2] list pertinent regulatory classification (eg RCRA — listed K011 waste, etc)

Figure 4.4 Example worksheet.

PROCESS AND WASTE CHARACTERIZATION

Firm _____	**Waste Minimization Assessment**	Prepared by _____
Site _____	Proc. Unit/Oper. _____	Checked by _____
Date _____	Project No _____	Sheet <u>4</u> of <u>4</u> Page__of__

| WORKSHEET **9d** | INDIVIDUAL WASTE STREAM CHARACTERIZATION | ♻ EPA |

(continued)

Waste stream _____

7. **Management method (continued)**

   Treatment
   - ☐ biological _____
   - ☐ oxidation/reduction _____
   - ☐ incineration _____
   - ☐ pH adjustment _____
   - ☐ precipitation _____
   - ☐ solidification _____
   - ☐ other (describe) _____

   - ☐ residue disposal/repository _____

   Final disposition
   - ☐ landfill _____
   - ☐ pond _____
   - ☐ lagoon _____
   - ☐ deep well _____
   - ☐ ocean _____
   - ☐ other (describe) _____

   Costs as of _____(quarter and year)

Cost element	Unit price $ per___	Reference/source
Onsite storage & handling		
Pretreatment		
Container		
Transportation fee		
Disposal fee		
Local taxes		
State tax		
Federal tax		
Total disposal cost		

Figure 4.5   Example worksheet.

Firm _____	**Waste Minimization Assessment**	Prepared by _____
Site _____	Proc. Unit/Oper. _____	Checked by _____
Date _____	Project No _____	Sheet 1 of 1 Page__of__

WORKSHEET 10   **WASTE STREAM SUMMARY**   **♻EPA**

Attribute	Description[1]		
	Stream no._____	Stream no._____	Stream no._____
Waste ID/name:			
Source/origin			
Component/or property of concern			
Annual generation rate (units____)			
Overall			
Component(s) of concern			
Cost of disposal			
Unit cost ($ per_____)			
Overall (per year)			
Method of Management[2]			

Priority rating criteria[3]	Relative Wt (W)	Rating (R)	R x W	Rating (R)	R x W	Rating (R)	R x W
Regulatory compliance							
Treatment/disposal cost							
Potential liability							
Waste quantity generated							
Waste hazard							
Safety hazard							
Minimization potential							
Potential to remove bottleneck							
Potential by-product recovery							
Sum of priority rating scores		$\sum$(R x W)		$\sum$(R x W)		$\sum$(R x W)	
Priority rank							

Notes: 1. Stream numbers, if applicable, should correspond to those used on process flow diagrams.
       2. For example, sanitary landfill, hazardous waste landfill, onsite recycle, incineration, combustion with heat recovery, distillation, dewatering, etc.
       3. Rate each stream in each category on a scale from 0 (none) to 10 (high).

Figure 4.6   Example worksheet.

Process surveys are notoriously difficult to perform and time consuming. The plant instrumentation is often not appropriate as it is aimed at control relative to some set point rather than at a quantitative measurement. The key elements of a survey are representative sampling and appropriate analysis. It is important to consider sampling of individual and combined streams as it may be better to treat individual streams separately rather than in combination. Care should also be taken to ensure good in-line mixing prior to sampling if any stream could be prone to segregation/separation — for example, if suspended solids are present, the duration of the survey and frequency of sampling depend on whether the process is continuous or batch and the likely stream variations; always err on the conservative side. Selection of the analytical technique(s) also influences the size and handling of the samples. Beware of changes/deterioration in samples if there is a time delay between sampling and analysis. In some cases, on-site analysis must be performed. There are a number of mobile laboratory services which can supplement in-house resources. If resources and time are limited then the highest priority waste streams in terms of environmental hazard and impact should be considered first, as well as those with a high degree of uncertainty.

For liquid effluents, the survey includes the drainage system. There are specialist firms that can perform such surveys rapidly using laser theodolite systems. It may be necessary to use dye tracing where there are uncertainties over the layout, and if the condition of the drains is poor a TV survey may assist. Do not overlook condensate as it may be contaminated.

The above assessment and survey only considers the planned sources of waste. It is equally important to identify the potential for unplanned creation of waste or release of waste. This includes housekeeping, maintenance, fugitive emissions, leaks, spills, plant malfunctions and the consequences of accidents and major incidents — for example, a very large quantity of waste can be created following a fire and bringing it under control. The waste manager needs to work closely with the safety manager in drawing up contingency plans to deal with such waste. This highlights the benefits of integrated SHE management. Another area for particular attention is any sunken or underground storage tanks which in the past have been notorious for contamination of land and groundwater.

Any assessment/survey benefits considerably if the plant operators and personnel are involved. The aims of the exercise should be explained and hands-on experience recognized as completing the picture.

If someone who is not familiar or directly involved in the process can participate in the assessment/survey, the result may be a fresher and more challenging view of the operation.

A useful systematic guide with extensive check-lists by HASTAM[6] covers management of waste, emissions and discharges and can be used to aid the assessment/survey. It is also available as PC software.

4.4 CHARACTERIZATION AND ANALYSIS

The assessment and survey described in Section 4.3 hopefully will have identified all of the waste streams. The characterization of each waste stream in terms of its composition and hazard is now considered. Table 4.5 lists the main characterization parameters and the associated analytical tests which may need to be carried out.

When requesting a characterization or analysis of a sample, it is important to define clearly what is required. General descriptions such as 'full analysis', 'hazardous components' or 'organics' leave considerable scope for misunderstandings. Garlick[9] has reviewed laboratory techniques and experiences in their application. In many cases, particularly where organics are involved, the sheer cost and time needed for detailed analysis may be prohibitively high unless it can be justified on sound consideration of risk.

4.5 MONITORING

The role of monitoring is to ensure continuous compliance under increasingly stringent legislation and to alert operators to developing or actual problems. It is also an integral part of environmental standards such as BS 7750[10]. To design the monitoring system so that it is respected by all as a valuable tool rather than a chore is important. If large volumes of monitoring data are generated but rarely referred to, then the results of monitoring will be prone to neglect. Choose the key characteristics with care — for example, there may be one parameter, such as pH, which is indicative of the overall performance of a treatment system. Too simplistic an approach, however, can risk problems being undetected until they become severe. The collected data may need to be presented in different formats for operators, engineers, managers, regulators and even the public. It is also important to classify the data according to its importance for immediate consideration and for longer term archiving.

If the treatment plant is automated, including the monitoring system, then there should be sufficient cross-checks to ensure that any problems which the automation may be masking are spotted early.

There are two basic types of monitoring:
- sampling and off-line analysis — often called spot or grab sampling and is the most common approach. On existing plant, it is generally already in use

TABLE 4.5
Characterization parameters and techniques

Parameter	Analytical technique and comments
Density	Also specific gravity (SG), useful in relating volume of waste to its mass. For liquids, simple ratio of weight of SG bottle with wastewater and distilled water corrected to 4°C. For solids, bulk density is the most important parameter. Solids compressed density may be useful if compaction is being considered.
Solids content/Total Suspended Solids (TSS)	Evaporation to dryness. Otherwise, separation by filtration or centrifugation, which will be method dependent.
Solids particle size	Vast range of definitions of size and size distribution with a large number of methods available. A major text is Terry Allen's book[7]. There may be significant differences between in-plant characteristics and results from laboratory tests.
Dewpoint	Temperature at which condensation of component(s) occurs on cooling a gas stream — measure of solvent content in gas stream. It can be used either in design of condensers or in identifying condensation problems in gas cleaning equipment. Acid dewpoints can be very high, causing severe corrosion and environmental problems.
Solubility	Mass (kg or kmol m^{-3}) of a substance dissolved in a solvent. It can be highly temperature dependent. An important parameter in assessing if a substance can be removed by filtration directly or separated by crystallization.
Element/compound composition	Vast range of analytical methods and instruments for inorganic and organic tests are available. Clearly, it is vital to identify likely components and limit tests to those critical to man and the environment. Table 4.6, page 82, lists the most significant techniques.

Continued overleaf

TABLE 4.5 (continued)
Characterization parameters and techniques

Parameter	Analytical technique and comments
Turbidity	Clarity of liquid from combined effect of any suspended solids and liquid properties determined from reduction in transmitted light. Suggested as an alternative to suspended solids parameter but difficult to correlate when particle size distribution and density in stream vary widely.
Colour	Usually ignored unless it is visible in discharges to air or aqueous environments. In water, it is affected by turbidity whilst in air it is affected by particulates/mist.
pH	Measure of acidity (low pH)/alkalinity (high pH) of water (pH = 7 is neutral) from electromotive force in a cell. A more correct measure of acidity/alkalinity is by a titration test. pH is significant for survival of living organisms and corrosion.
Chemical oxygen demand (COD)	Wastewater test based on a strong oxidizing agent, potassium dichromate; 2 h to complete; COD > BOD; a high COD:BOD ratio indicates hard to biodegrade organics and/or toxic chemicals
Biological/ biochemical oxygen demand (BOD)	Wastewater test based on incubation for 5 days, expressed as mg of oxygen needed to oxidize organics in 1 litre of water sample. It is used to indicate level of sewage pollution, only biodegradable organics measured, and prone to error from poisoning of bacteria.
Total organic carbon (TOC)	Suggested alternative to BOD using either high temperature catalytic oxidation or UV based oxidation but can give differing results for some samples, so needs careful interpretation as it is method dependent
Listed substance	See Table 3.4, page 30 Classed as special waste
Prescription-only medicine	Classed as special waste
Conductivity	Indicates level of electrolytes/ions present, standard conductivity cells, temperature dependent

Continued opposite

TABLE 4.5 (continued)
Characterization parameters and techniques

Parameter	Analytical technique and comments
Flash point	Lowest temperature for ignition of vapour observed by progressively raising the temperature of a closed sample until it can be ignited
Calorific value	BS 7420[8] for fuels can be used and is based on bomb calorimetry
Oil in water	UV/microwave absorption and UV fluorescence
Total VOCs in water	Headspace analysis or photo-ionization and flame ionization
Toxicity	In most cases, lists of toxic substances are published and exposure limits set. However, for complex mixtures it may be necessary to perform a Direct Toxicity Assessment (DTA) using living organisms.
Corrosivity	In addition to pH, corrosion which affects the surface of solids can be due to galvanic corrosion in metals and sulphide corrosion in concrete. The lack of corrosiveness in waters is measured by the Langelier saturation index which determines the over- or under- saturation of calcium carbonate which can form a protective surface film.
Biodegradability	Originally focused on surfactants which cause foaming problems in wastewaters and organic matter in landfills which produce off-gases, a variety of methods depending on whether primary or ultimate degradation is being assessed.
Carcinogenicity	Some carcinogens are well known and exposure limits defined. However, many organic, chloro-organic and even inorganic compounds may have a contributory effect on the development of certain cancers. Virtually nothing is known about the carcinogenic effects of combinations of substances (see Section 8.8, page 178).
Radioactivity	This can be a complex topic and expert advice should be sought if there is concern that radioactive substances may be present in a waste. Simple total radiation measurements are not suitable for characterization.

TABLE 4.6
Composition analytical techniques[9]

Component	Technique
Metals	• atomic absorption • atomic fluorescence • atomic emission • mass spectrometry • X-ray fluorescence • polarography • voltametry • colorimetry • ion chromatography
Anions	• colorimetry • ion chromatography • X-ray fluorescence • titrimetry • gravimetry
Organics	• gas chromatography • liquid chromatography • mass spectrometry • infrared spectrometry • ultraviolet spectrometry • colorimetry
Other components	• amperometry • gravimetry • nephelometry (photometric technique)

when the decision to move to continuous monitoring is taken and it also provides the data on which the continuous monitoring system is selected and designed. For routine monitoring, however, it suffers from the possible disadvantage of missing significant excursions in the value of the parameters being monitored if the sampling is either over too short or too long a period. Also, errors can be introduced during the handling of samples, which requires high levels of QA and hence human resources.

• continuous on-line analysis — also called *in-situ* systems. It offers major potential to track the precise performance in terms of variations and averages. Currently, there is a limited number of continuous monitors available and their maintenance and calibration needs can be high.

There is a major difference between gaseous and liquid monitoring. For liquid/aqueous streams, there are a number of well established test standards and methodologies as listed in Table 4.7. For gases, however, there are few practicable standards and the most widely recognized approach has been set down by the US EPA. Averdieck[14] has reviewed isokinetic gaseous emissions monitoring to BS 3405[13].

The pace of technical development in continuous monitoring is accelerating in this extremely active field. Over recent years, one of the fastest growing areas has been the monitoring of volatile organic compounds. Future trends will see major developments in:
- monitoring and control of odours;
- biological controls;
- real-time dioxin monitoring;
- toxicity-based monitoring — for example, biological probes.

TABLE 4.7
Standard test/analytical methods

Wastewaters	• UK Department of the Environment's Standing Committee of Analysts, *Methods for the Examination of Waters and Associated Materials* (HMSO, London, UK) • *Standard Methods for the Examination of Water and Wastewater* (American Public Health Association, American Waterworks Association and the Water Pollution Control Federation, Washington, USA) • Methods for Organic Chemical Analysis of Municipal and Industrial Wastewater (US EPA, EMSL, Cincinnati, USA) • BS 6068 series: water quality[11]
Air/gaseous wastes	• *Perry's Chemical Engineer's Handbook* — section 26–31 — covers emissions measurement and describes sampling and standard US EPA methods for flowrate, molecular weight, CO_2 and O_2 content, humidity, particulates, SO_2, NO_x, CO, fluorides, organics and hydrocarbons and plume opacity • US EPA methods 'Code of Federal Regulations', volume 40, pts 53–60 • BS 6069 series: characterization of air quality[12] • BS 3405: method for measurement of particulate emission including grit and dust (simplified method)[13]

4.6 WASTE DATABASE

The importance of retaining and collating information on waste, effluents and emissions for a variety of purposes and audiences necessitates a well organized and managed database. The database includes the following information:

- a general description of the waste stream and its origin;
- the flow/quantity generated (performance versus consent);
- the chemical composition and the concentration of the components (performance versus consent);
- the storage area(s) on the site;
- movements into and out of storage and amounts currently in stock;
- a description of the container(s) used and appropriate marks and labels;
- any hazards associated with the waste;
- routine handling procedures including the need for any precautionary measures;
- emergency procedures;
- details of consignments leaving the site and their final destination. In the UK it is a statutory requirement to keep a copy of the consignment note which accompanies special waste movements;
- monitoring data from on-site treatment of gaseous wastes;
- monitoring data from on-site treatment of liquid wastes.

The records should be kept in an orderly and accessible manner, but the exact mode of retention — for example, electronic or hard copy — will depend on the particular company circumstances. It is essential that information is kept up-to-date and trends monitored. Any significant divergence from the norm needs to be investigated.

There are several commercial computer programs which are specifically aimed at waste management and the most up-to-date releases are listed in the annual *CEP Software Directory*[15].

4.7 MARKING AND LABELLING

If the waste is packaged then the appropriate marking and labelling needs to be used. There are regulations covering the transport of dangerous goods which may apply to the waste. *Croner's Waste Management Guide*[3] has a comprehensive section of 'waste charts' which set out for each waste material/compound details on labelling. A more comprehensive guide to the transport of dangerous goods has been produced by Castle[16]. Further details on off-site issues are covered in Chapter 7.

REFERENCES IN CHAPTER 4

1. *Kirk-Othmer Encyclopedia of Chemical Technology*, 4th edition (Wiley-Interscience, Chichester, UK).
2. Austin, G.T., 1984, *Shreve's Chemical Process Industries*, 5th edition (McGraw Hill, New York, USA).
3. *Croner's Waste Management Guide*, 1991 (Croner Publications, UK).
4. US Environmental Protection Agency, 1988, *Waste Minimisation Opportunity Assessment Manual* (US EPA Hazardous Waste Research Laboratory, Office of R&D, Cincinnati, USA).
5. US Environmental Protection Agency, 1991, *Draft Guide for an Effective Pollution Prevention Program* (US EPA Hazardous Waste Research Laboratory, Office of R&D, Cininnati, USA).
6. HASTAM, 1991, *Environmental Audits* (Mercury Books, London, UK).
7. Allen, T., 1981, *Particle Size Measurement* (Chapman and Hall, London, UK).
8. *BS 7420, Guide for determination of calorific values of solid, liquid and gaseous fuels (including definitions)*, 1991 (British Standards Institution, UK).
9. Garlick, J.R., 1993, Laboratory analytical techniques for materials of environmental significance, *IChemE Symp Ser 132, Effluent Treatment and Waste Minimization*, 123–129.
10. *BS 7750: Specification for environmental management systems*, 1994 (British Standards Institution, UK)
11. *BS 6068 Parts 0 to 6 series: Water Quality* (British Standards Institution, UK).
12. *BS 6069 Parts 1 to 5 series: Characterization of air quality* (British Standards Institution, UK).
13. *BS 3405: Method for measurement of particulate emission including grit and dust (simplified method)*, 1983 (1989) (British Standards Institution, UK).
14. Averdieck, W., 1995, Emission monitoring: hints and tips on EPA compliance, *Environ Man J*, 3 (3): 15–16.
15. *CEP Software Directory*, annual, a supplement to Chemical Engineering Progress (AIChE, New York, USA).
16. Castle, M., 1995, *The Transport of Dangerous Goods: A Short Guide to the International Regulations*, 2nd edition (Pira International, Leatherhead, UK).

5. WASTE MINIMIZATION

5.1 INTRODUCTION

Liquid, solid and gaseous waste materials are inevitably generated during the manufacture of any product. Apart from creating potential environment problems, wastes not only represent losses from the production process of valuable raw materials and energy, but also require significant investment in pollution control practices. In effect, producers pay for their waste twice — first in 'lost' product and second in disposal. Industrial waste treatment is generally viewed as an addition to the end of a process, offering little scope to recover value from the waste material. Worse still, many 'end-of-pipe' waste treatment techniques do not actually eliminate the waste but simply transfer it from one environmental medium to another — that is, air, water and land — often in a highly diluted form.

The UK Environmental Protection Act 1990 (EPA 1990) now requires a continuing reappraisal by the process industries of waste management practices, as improved technology becomes available and regulatory requirements are progressively introduced. It is likely that, as further restrictions are placed on the disposal of substances to environmental media, the costs of waste treatment and disposal will rise. Against this background, there will be incentives to minimize the generation of all forms of waste, not only for environmental protection but also for commercial benefit. Table 5.1, page 88, gives an overview of Chapter 5.

Whilst this chapter mainly concentrates on existing processes, many of the principles can nonetheless be applied to the design of new processes. In general, it is better to use the design, research and development stages to avoid waste generation than to modify a process once it has been installed. Such 'clean processes' or 'clean technologies' should not need to rely substantially on end-of-pipe pollution abatement techniques to reduce impact on the environment.

In general, the process industries have devoted a considerable amount of effort to research and development with a view to maximizing product yields. This has been achieved by minimizing by-products and waste in order to reduce raw material costs, which are often a high proportion of the total cost. More emphasis was put on this with the implementation of the UK Control of Pollution Act and the system of charging for liquid effluents based on volume, COD

TABLE 5.1
Overview of Chapter 5

5.2, page 89 Organization	• company policy and responsibilities of management and employees
5.3, page 90 Basic approach	• project outline and key steps • mass balances • batch and semi-batch processes • water conservation
5.4, page 92 Techniques	• good operating practices • technological changes • input material changes • product changes • recycling
5.5, page 97 Priorities and targets	• factors driving priority • setting of target levels
5.6, page 98 Reviews, audits and corrections	• updating minimization programmes • regular environmental auditing
5.7, page 98 The club approach	• industrial collaboration to share knowledge

and % solids. But minor emissions to drain or atmosphere and small losses of solids have not received so much attention. A waste minimization approach changes this emphasis by studying the whole process from beginning to end.

By implementing a waste minimization policy, savings can be expected through reduced on-site waste monitoring, control and treatment costs and through reduced handling, pretreatment, transport and off-site disposal costs. Reduced storage, administration, analysis and production costs which include raw materials, energy and utility requirements can also add to company savings. In addition, there are reduced risks from handling hazardous materials resulting in better health and safety, and reduced risks to the environment which could be manifested by reductions in, or elimination of, liability charges. From a public relations viewpoint, the company's image in the eyes of its shareholders, employees and the local community is enhanced.

Waste minimization projects are evaluated in the same manner as any other business opportunity. The implementation of a waste minimization project is likely to incur additional capital investment. It is therefore important that all potential benefits are correctly appraised and quantified, although this may be difficult for some of the less tangible benefits. As far as possible, the true and total costs of waste management should be included in the analysis for each specific project.

5.2   ORGANIZATION

The IChemE's *Waste Minimization: A Practical Guide*[1] provides a methodology for implementing waste minimization policies, programmes and projects. It is essential that the senior management within a company provides the lead through a policy commitment to waste minimization. Its overall objectives, strategies and timescales for their achievement are defined in the organization plan for the implementation of its environmental management system — for example, British Standard on Environmental Management Systems[2], the Chemical Industries Association Guidelines on Responsible Care[3] and the European Community Eco-Audit and Management Regulation[4].

A practical commitment is required to make the difference between simply preparing an overall objective, such as that given in the company's policy statement, and preparing a specific plan which can be successfully implemented, audited and reviewed. A senior manager may need to be allocated the responsibility for waste minimization programmes. In small companies, this person may be given the responsibility on top of other duties, perhaps combined with safety and welfare, or a line function like purchasing or warehousing. It may not be the best policy to give this task to a manager with production responsibilities because of an obvious conflict of interests. In large companies, a dedicated department of specialists could be justifiable.

The execution of a waste minimization policy becomes the responsibility of the various line managers engaged in carrying out the waste minimization projects and procedures in their particular departments. The management of waste minimization must be seen as an integral part of every employee's job in much the same way as safety. An effective way of achieving this is to allocate waste disposal costs to the department in which the waste arises, providing that the allocation has a reasonably justifiable basis. There is little point in setting up a pragmatic system which simply shuffles costs between departments. Nor is there any point in allocating insignificant costs. Additional managerial responsibilities include:
* ensuring that economic, technical, regulatory and institutional barriers to progress are overcome (Reference 1 provides more information);
* providing employees with training and motivation;

- monitoring of progress and assisting with feedback;
- ensuring that the overall programme allows lower priority projects to be ultimately implemented.

5.3 BASIC APPROACH

The basic methodology of waste minimization detailed in the IChemE's *Waste Minimization Guide*[1] is summarized in Figure 5.1, page 91. A possible outline of a typical waste minimization project would be to:
- identify and quantify all sources of waste, including solid waste, spills and discharges to atmosphere and effluent;
- consider what waste reduction and pollution prevention options are available;
- examine process routes and engineering practices to select those which give the best options, from economic, technical and environmental viewpoints;
- agree options and prioritize;
- set targets for the reduction or elimination of waste streams in terms such as volume, mass, concentration and frequency of discharge;
- define and allocate appropriate resources;
- implement agreed actions;
- monitor and record results and review progress against the agreed targets;
- take corrective action, as necessary.

The IChemE guide provides much greater detail on:
- setting goals and timescales for achievement;
- creating an assessment and evaluation team;
- carrying out the assessment phase which includes how data can be acquired and interpreted;
- carrying out a site review;
- ranking of practical options by technical and economic evaluations;
- reporting on the formal assessments and evaluations;
- implementing the best options including methods to review progress and audit performance.

A critical element of the assessment phase is the mass balance which accounts for all inputs, including impurities in raw materials and feedstocks, and all outputs whether finished products, by-products, gaseous emissions, waste liquids or solids. Minor components in waste streams — for example, products or unreacted feedstocks entrained in an aqueous distillate which might normally go direct to drain as harmless — must be identified and included. Although such components may be environmentally harmless, they may make a considerable contribution to COD and therefore cost.

WASTE MINIMIZATION

```
┌─────────────────────────────────┐
│ Company policy and strategy for │
│ its implementation through waste│
│ minimization                    │
└─────────────────────────────────┘

                                    ┌─────────────────────────────────────┐
                                    │ Senior management commitment        │
                    Re-evaluate     │                                     │
                    overall goals   │ Set goals and time-scales           │
                                    │                                     │
                                    │ Establish assessment and evaluation │
                                    │ team                                │
                                    └─────────────────────────────────────┘

┌─────────────────────────────────────────┐
│ Assessment phase                        │
│   ● Data collection                     │
│   ● Organization of data                │     Lower priority
│   ● Identification of significant waste │     projects
│     generation practices                │
│   ● Site review                         │
└─────────────────────────────────────────┘

                                    ┌─────────────────────────────────┐
                                    │ Preliminary ranking of options  │
                                    └─────────────────────────────────┘

┌─────────────────────────────────────────┐
│ Feasibility analysis phase              │
│   ● Technical evaluation                │
│   ● Economic evaluation                 │
└─────────────────────────────────────────┘

                                    ┌─────────────────────────────────┐
                                    │ Report on assessment and        │
                                    │ evaluation                      │
                                    │                                 │
                                    │ Implementation of waste         │
                                    │ minimization projects           │
                                    │                                 │
                                    │ Review and audit of waste       │
                                    │ minimization projects           │
                                    └─────────────────────────────────┘

┌─────────────────────────┐
│ Feedback information    │
└─────────────────────────┘
```

Figure 5.1 Methodology of waste minimization.

The IChemE's *Waste Minimization Guide*[1] addresses in detail the problems associated with establishing an accurate mass balance which includes all resources, products and wastes. The parameters which need to be identified for each stream are:
- total mass or volume;
- maximum, minimum and average flowrates;
- maximum, minimum and average concentrations;
- concentrations under unsteady conditions such as start-up, shutdown, equipment cleaning and critical equipment failure.

Special care needs to be taken for batch and semi-batch processes carried out individually or in campaigns. Important areas for waste minimization also lie in the provision of site utilities, transport and office needs. In this respect, it is important to recognize that significant quantities of waste can be generated by kitchens, stores and offices which support the processing facility. Energy conservation, a well established practice in the process industries, often reduces pollution and waste generation. For example:
- a reduction in energy demand decreases the quantity of fossil fuels burned, thereby decreasing the quantity of air pollutants generated;
- a reduction in steam demand decreases the discharge of cooling water and boiler blowdowns;
- a reduction in the amount of boiler feed water requires less use of chemicals for treatment purposes.

In addition, since waste minimization projects are concerned with reducing the total use of resources, so their implementation should also result in a more efficient usage of energy.

Water conservation is becoming increasingly more important as demands grow for higher consumption at higher quality. The waste minimization techniques described in this chapter can be applied directly to water abstraction, use, cooling and recycling systems. Extend water conservation to all activities on the process site, including kitchens, toilets and washing facilities. Implement programmes to obtain minimum fuel consumption for the transport fleet and to minimize the generation of office waste, particularly paper. Encourage recycling of paper and plastic waste from offices.

5.4 TECHNIQUES

The IChemE's *Waste Minimization Guide*[1] describes the hierarchy of waste management practices and the roles of waste minimization and recycling within it. It also describes some of the 'toolkits' which are available to process, plant and design engineers to help the selection of good waste minimization projects.

The complete prevention of waste is a laudable objective but, perhaps with few exceptions, an unachievable one. The next best objective is to reduce waste generation at source by as much as possible through the implementation of:
- good operating practices;
- technological changes;
- input material changes;
- product changes.

Recycling offers considerable scope for reducing environmental impacts but implies that valuable resources are being processed at higher than minimum flowrates. Equipment sizes and energy demands are strongly related to recycle flows and, as a consequence, recycling either on or off site, involving reuse, use or reclamation techniques, should be lower down the hierarchy of good waste management practices.

5.4.1 GOOD OPERATING PRACTICES

Good operating practices, housekeeping, engineering and maintenance, which use operational improvements or administrative changes to reduce the generation of waste material, can often be implemented relatively quickly at minimal capital cost and with short payback. Examples for raw materials, products and wastes include:
- clear specification of good housekeeping and materials handling procedures;
- implementing quality assurance techniques including control of raw material quality and making the product 'right first time';
- regular auditing of materials purchased against materials used;
- avoidance of over-ordering;
- regular preventive maintenance;
- segregation of waste streams to avoid cross-contamination of hazardous and non-hazardous materials, and to increase recoverability;
- reduction in the volume of wastes by filtration, membrane processes, vaporization, drying and compaction;
- proper fitting lids and vapour traps to solvent tanks;
- avoidance of corrosion, including drums and containers;
- elimination of poor storage conditions;
- proper labelling of containers;
- improvement of maintenance scheduling, record-keeping and procedures to increase efficiency;
- re-evaluation of shelf-life characteristics to avoid unnecessary disposal of long-life materials;
- improvement of inventory and management control procedures;

- changes from small volume containers to bulk or reusable containers;
- introduction of employee training and motivation schemes for waste reduction;
- collection of spilled or leaked material for reuse;
- consolidation of types of chemicals to reduce quantities and types of wastes;
- rescheduling of production to reduce frequency and number of equipment cleaning operations.

It is important that proper attention be given to eliminating or minimizing spills, leaks and contamination during the storage of raw materials, products and process wastes, and the transfer of these materials within the production facility. Examples requiring attention include:
- leaking valves, hoses, pipes and pumps;
- leaking tanks and punctured containers;
- overheating of tanks and drums;
- overfilling of tanks;
- inadequate, poorly maintained or malfunctioning high level protection;
- gas generation in drums;
- leaks and spills during material transfer;
- inadequate bunding;
- leaking filters, bunkers and bins in powder transfer operations;
- equipment and tank cleaning operations;
- contamination to produce off-specification raw materials and products by inadequate process control or by the entry of adulterating substances;
- lack of regular maintenance, inspection and operator training;
- incorrect sequencing of valve operations.

Sampling systems may deserve special attention:
- consider if a sample is essential — often process parameters determine composition and samples should be limited to those which continue to give useful data;
- minimize the sample size to reduce laboratory waste;
- design sample points so that forerunnings can be eliminated or recycled wherever possible;
- sample points should be as close to the vessel or line as possible and fitted with as small a valve as is practicable.

5.4.2 TECHNOLOGICAL CHANGES

Technology changes concern process and equipment modifications in order to reduce waste primarily within the production environment. The modifications may involve the use of new or modified processes and hardware to lessen or prevent pollution. Examples include:

- introduction of new processes or equipment which produce less waste — that is, clean technologies;
- fundamental changes to or better control of process operating conditions such as flowrate, temperature, pressure, residence time and stoichiometry to reduce waste and consume less raw materials and energy;
- redesign of equipment and piping to reduce the amount of materials to be disposed of during start-ups, shutdowns, product changes and maintenance programmes;
- installation of vapour recovery systems and/or vent balancing to return emissions to the process;
- changes to mechanical cleaning to avoid the use of solvents and the generation of dilute liquid wastes, provided such changes are not otherwise detrimental;
- use of more efficient motors and speed control systems to reduce energy consumption.

Such retrofit operations employ current technologies and it is inevitable that a further stage of waste minimization activity will evolve. Of much greater significance in the future will be the introduction of new or redesigned industrial processes which are inherently less polluting — that is, the 'clean' technologies — which will involve highly selective separation and reaction technologies specifically designed for waste minimization.

Design tools are now being developed to aid waste minimization via technological changes. Specific methods are outlined in the IChemE guide[1].

The introduction of a technological change normally requires a certain amount of capital investment and may therefore incur greater resistance to implementation. This is especially true if it is perceived that there is some risk involved. In order to overcome this barrier, it is essential that all of the less tangible and less easily quantifiable benefits are incorporated into the profitability analysis.

5.4.3 INPUT MATERIAL CHANGES

Hazardous materials used in a production process — for example, raw materials, solvents, catalyst supports and so on — may be replaceable by less hazardous or even non-hazardous materials. Changes in input materials may also lead to a reduction in, or avoidance of, the formation of hazardous substances. The objective also includes a reduction in the quantity of waste generated. Some examples are:

- replacement of chlorinated solvents by non-chlorinated solvents, water or alkaline solutions in cleaning and degreasing operations;
- substitution of chemical biocides by alternatives, such as ozone;

- replacement of solvent-based paint, ink and adhesive formulations with water-based materials or UV/electron beam cured formulations which are effectively 100% solids;
- substitution of a more durable coating to increase coating life;
- increase in the purity of purchased raw materials to eliminate the use of trace quantities of hazardous impurities;
- reduction of phosphorus in wastewater by reduction in the use of phosphate-containing chemicals;
- replacement of hexavalent chromium salts by trivalent chromium salts in plating applications;
- replacement of solvent-based by water-based developing systems in printed circuit-board manufacture;
- replacement of cyanide plating baths with less toxic alternatives.

One possible problem with material changes is that they might have an adverse effect on the production process, product quality and even waste generation — for example, changing from a solvent-based to a water-based product might increase wastewater volumes and concentrations, leading to increased wastewater treatment and sludge disposal costs. Clearly, all of the possible impacts of a proposed change must be evaluated.

5.4.4 PRODUCT CHANGES

Product changes are reformulations of final or intermediate products, performed by the manufacturer in order to reduce the quantity of waste arising from their manufacture. Other objectives might include:
- a change in a product's specification in order to reduce the quantity of chemicals used;
- a modification of the composition or final form of a product to make it environmentally benign;
- changes to reduce or modify packaging.

Product reformulation is not particularly easy. Increasingly it will become necessary to carry out life cycle analyses in order to determine the change in overall impact on the environment and not just within the process boundary. One of the main causes of resistance is changing the customer's process to handle the change. The need to test, trial and change production may be dictated by the need to have a compensating cost/benefit incentive.

Clearly, it is important to recognize the need to retain technical and economic objectivity. Be pragmatic. In particular, aim for a good balance between environmental and commercial advantages.

5.4.5 RECYCLING

Recycling waste material for reuse, use and reclamation may in many circumstances provide a cost-effective alternative to treatment and disposal.

Success depends on:
- the ability to reuse waste material by return to the originating process as a substitute for input material;
- the ability to use waste material as a raw material either on site or off site;
- the ability to segregate recoverable and valuable materials from a waste which may be mixed and is of generally low value (reclamation).

The best place to recover wastes is within the production facility and the following are good candidates for recycling:
- contaminated versions of process raw materials can be used to reduce raw material purchases and waste disposal costs — such waste can be recovered at the point of source but some purification might be required;
- lightly contaminated wastes that can be used in other operations which do not require high purity materials;
- wastes which have physical and chemical properties suitable for other on-site applications— for example, the use of a caustic waste stream to neutralize an acid waste stream or the use of waste solvents, oil and so on, in combustion processes;
- reuse of extracted water from dilute, high volume waste streams;
- wastes which can be refined on site, either in the main process — for example, the recycling of slop oils in an oil refinery — or in a special purpose plant — for example, a solvent refining unit.

Wastes may be considered for use or reclamation offsite when:
- equipment is not available on site to do the job;
- not enough waste is generated to make on-site recycling cost-effective;
- the recovered material cannot be used in the production process.

Materials commonly reprocessed off site by chemical and physical methods include oils, solvents, electroplating wastes, lead-acid batteries, scrap metal, food processing waste, plastic waste and cardboard. Some wastes have a use without the need for reprocessing or refining — for example, waste acids and alkalis.

5.5 PRIORITIES AND TARGETS

Having identified and quantified all of the waste streams and considered the techniques available for dealing with them, the next step is to decide the priority in which they are managed. Technical and economic evaluations aid the ranking process but priority may be driven by needs and desires to:

MANAGEMENT OF PROCESS INDUSTRY WASTE

- achieve compliance;
- provide a bigger margin between actual performance and compliance (this may be to eliminate occasional excursions beyond compliance levels, to anticipate stricter future standards or to create financial or other commercial opportunities);
- respond to complaints (public perception may not always be logical or valid but it should be dealt with constructively);
- anticipate future changes in availability and cost of existing disposal routes;
- make general improvements as a matter of company policy or continual improvement.

For each of the wastes identified for action, a target level must be set and subsequently monitored. The targets are used in the specifications for design and performance, especially when technological changes need to be implemented.

5.6 REVIEWS, AUDITS AND CORRECTIONS

The process industries are continually changing by developing new feedstocks, products, processes, equipment designs, analytical techniques and so on. A waste minimization plan needs to take advantage of these developments since their impact on the nature, composition, volume and cost of disposal must be assessed for each waste. These factors are considered in reviews of the waste minimization programme and changes are made to it, as required.

The effectiveness of waste minimization procedures and their implementation are assessed by regular environmental audits made by trained auditors. Provided that internal audits are carried out conscientiously and by staff not involved in the department being audited, an external audit may be necessary only once in every three or four years in order to obtain an independent view. Audits should not be regarded as a means for policing a system; they are constructive and provide ways of improving the system and its procedures. Shortcomings or discrepancies identified by an audit are corrected against an agreed timescale, usually under the supervision of the managers of the department in which they occur. The IChemE training package on environmental auditing can used to guide the introduction of auditing procedures[5].

5.7 THE CLUB APPROACH

By sharing experiences in club-type activities, many of the barriers to implementing good engineering waste minimization projects described in the IChemE guide[1] can be broken down to create short-term commercial and environmental opportunities. Specific examples of such activities are the PRISMA project in The Netherlands, the Aire and Calder project and the regional clubs being established under the auspices of the UK Department of Trade and Industry. One important reason for having a regional club is the common need to protect the quality of the atmosphere and waterways around areas of significant processing activity.

REFERENCES IN CHAPTER 5
1. Crittenden, B. and Kolaczkowski, S., 1995, *Waste Minimization: A Practical Guide* (IChemE, Rugby, UK).
2. *BS 7750: Specification for environmental management systems*, 1994 (British Standards Institution, UK).
3. *Guidelines on Responsible Care*, 1996 (CIA, UK).
4. *European Community Eco-Audit and Management Regulation*, 1996, (HMSO, UK).
5. *Environmental Auditing Training Package*, 1993 (IChemE, Rugby, UK).

6. ON-SITE WASTE TREATMENT

6.1 INTRODUCTION

After minimizing waste as far as is technically and economically possible, the process could still produce some residual material which may have to be treated either on or off the site before it can be safely discharged to the environment. The decision to treat or not is directly related to the hazard to people and the environment. The options to consider are:
- discharge/off-site disposal;
- on-site treatment and discharge;
- on-site treatment and off-site disposal;
- off-site treatment and disposal.

Discharge to air or the aqueous environment may need consideration under Integrated Pollution Control. There is an increasing tendency towards on-site treatment of hazardous wastes, or where there may be particular waste handling/disposal problems. This is reducing the need for off-site disposal of such materials. Clearly, the decision on the best practicable environmental option (BPEO) must be based on a sound assessment of the risks and the economics involved. This chapter only considers the on-site options (see Table 6.1, page 102). Off-site disposal is covered in Chapter 7; note that incineration is discussed in Chapter 7 but can be used as an effective on-site option. When considering on-site treatment, there are two broad options:
- at-source treatment (keeping wastes segregated);
- centralized treatment (mixing of wastes).

The term 'end-of-pipe' technology has been widely applied to environmental protection and implies adding onto the process some kind of treatment device(s). It has attracted a negative connotation in recent years as taking the easy option rather than fully evaluating the options to reduce, recycle or recover the waste stream. However, in many cases such end-of-pipe methods may well be the most effective choice both technically and economically.

If a new process has been conceived, or is being demonstrated, then there may be grants or awards from Government or EU programmes. The UK Engineering and Physical Sciences Research Council (EPSRC) includes a clean technology programme and funds university research in which industry is encouraged to collaborate.

TABLE 6.1
Overview of Chapter 6

6.2, page 102 Selection of treatment methods	• principal steps in selecting BPEO and treatment technologies
6.3, page 105 Gaseous treatment technologies	• review of unit operations for separation/destruction of particulates, droplets and other components from gases • their advantages and disadvantages • selection guide
6.4, page 124 Liquid treatment technologies	• review of unit operations for separation/destruction of particulates and soluble components from liquids • their advantages and disadvantages • selection guide
6.5, page 134 Solids treatment technologies	• review of unit operations for stabilization/conversion/destruction of solids • their advantages and disadvantages • selection guide
6.6, page 135 Incineration	• principles of combustion • incinerator technology options • waste-to-energy • pyrolysis • cement kilns

For a new in-house waste treatment facility, the importance of public acceptability cannot be overstated. In the event that substantial resistance to a proposed facility arises, there could be significant cost implications associated with appeal procedures. Corporate damage could also be sustained due to adverse public reaction.

6.2 SELECTION OF TREATMENT METHODS

An overview of the stages in the selection of the BPEO proposed by the UK Chemical Industries Association (CIA) is given in Table 6.2.

The principle of BPEO is now well established but the practice of selection of BPEO is still evolving. It is not possible to define BPEO in simple terms which can be applied generally, and therefore a body of experience is being built up both in the industrial community and Government organizations.

TABLE 6.2
Selecting a best practical environmental option (BPEO)

Step	Action	Comment
1	Define the objective	State the objective of the project or proposal at the outset, in terms which do not prejudge the means by which that objective is to be achieved
2	Generate options	Identify all feasible options for achieving the objective: the aim is to find those which are both practicable and environmentally acceptable
3	Evaluate the options	Analyse these options, particularly to expose advantages and disadvantages for the environment. Use quantitative methods when these are appropriate. Qualitative evaluation will also be needed.
4	Summarize and present the evaluation	Present the results of the evaluation concisely and objectively, and in a format which can highlight the advantages and disadvantages of each option. Do not combine the results of different measurements and forecasts if this would obscure information which is important to the decision.
5	Select the preferred option	Select the BPEO from the feasible options. The choice will depend on the weight given to the environmental impacts and associated risks, and to the costs involved. Decision-makers should be able to demonstrate that the preferred option does not involve unacceptable consequences for the environment.
6	Review the preferred option	Scrutinize closely the proposed detailed design and the operating procedures to ensure that no pollution risks or hazards have been overlooked. It is good practice to have the scrutiny done by individuals who are independent of the original team.

Continued overleaf

TABLE 6.2 (continued)
Selecting a best practical environmental option (BPEO)

Step	Action	Comment
7	Implement and monitor	Monitor the achieved performance against the desired targets especially those for environmental quality. Do this to establish whether the assumptions in the design are correct and to provide feedback for future development of proposals and designs.
1–7	Maintain an audit trail	Record the basis for any choices or decisions through all of these stages — that is, the assumptions used, details of evaluation procedures, the reliability and origins of the data, the affiliation of those involved in the analytical work and a record of those taking the decisions.

Note — the boundaries between each of the steps will not always be clear-cut as some may proceed in parallel or need to be repeated.

It is advisable to utilize the regulatory authorities throughout the various stages of a project to ensure that decisions are based on the most up-to-date practice.

A wide range of techniques is available for treating a variety of gaseous, liquid and solid waste streams. In general, these techniques can be divided into physical, chemical, thermal and biological methods.

Croner's Waste Management Guide[1] has a comprehensive section of 'waste charts' which set out for each waste material/compound the treatment and disposal options. There are also notes on handling precautions, labelling, storage, legislation, physical, chemical, medical and environmental hazards.

In general, a combination of unit operations usually yields the best treatment solution. Some simple unit operations offer essentially pretreatment performance, whilst some more sophisticated devices are ideal for final treatment prior to discharge. A systems approach is essential to lead to a technical and economic comparison of various alternative schemes. Most unit operations separate the inlet stream into two fractions, one having a reduced concentration of the critical waste component and the other an increased concentration. The former stream may be sufficiently clean to be recycled or discharged to the environment. The latter stream requires either further treatment, recycle or disposal. Many waste treatment operations ultimately produce a solid, powder or sludge.

TABLE 6.3
Selection factors

Environmental	• space availability/location options • local environment • utilities availability • existing facilities • maximum emission limits • aesthetics (plumes, colours and odours) • additional source of noise
Engineering	• critical component characteristics • waste stream characteristics • equipment design and performance characteristics
Economic	• capital costs • operating costs • life cycle costs

In selecting a treatment system, give due regard to making it as flexible and 'future-proof' as possible against more stringent legislation and process changes. The principal criteria for initially selecting the feasible options and then deciding on the preferred option are listed in Table 6.3.

6.3 GASEOUS TREATMENT TECHNOLOGIES

Waste gas streams may contain particulates, droplets or gaseous components (organics and/or inorganics) singly or in combination which must be removed. The gaseous components can range from products of combustion to volatile organic compounds (VOCs) and to odours which can pose the most difficult treatment challenges. There is increasing concern over fine particulates as demonstrated by the debate over particles in diesel combustion exhausts. A future trend will be integrated systems treating several components — for example, NO_x and SO_x. Sittig[2] has compiled a comprehensive catalogue of particulate emission sources and treatment solutions classified by industry sector, which is useful to gain an appreciation of past industrial practice although more modern solutions may be available.

The wide range of devices available for removal of particulates and droplets is summarized in Figure 6.1, page 106–107. Table 6.4, page 108–113, lists the treatment options for particles, droplets and gaseous contaminants, and briefly covers their advantages and disadvantages. Figure 6.2, page 114–116,

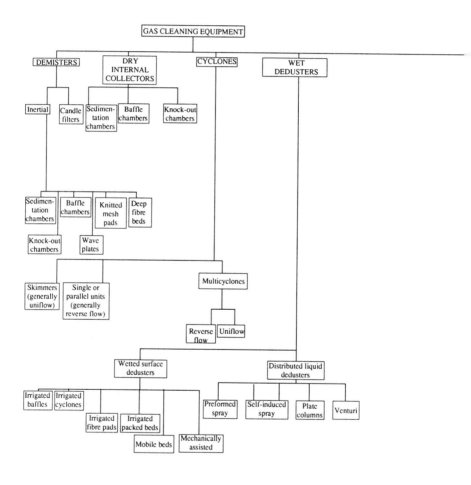

Figure 6.1 Classification of gas cleaning devices for particulates and droplets.

ON-SITE WASTE TREATMENT

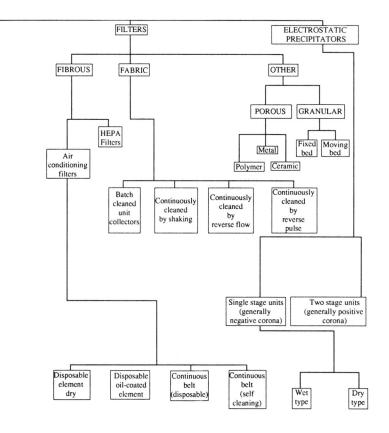

TABLE 6.4
Gaseous treatment technologies

Waste component	Technique	Advantages
Particulates	Cyclones, centrifugal separators	• simple • cheap • compact • low pressure drop • high temperature operation Note — multicyclones may offer better performance but have increased cost, pressure drop and are prone to design and operating problems
	Bag filters	• high efficiency to submicron particles • tolerant to operating variations
	Electrostatic precipitators	• very efficient • low pressure drop • suitable for high temperature and pressure • good for flowrates
	Wet scrubbers	• compact • combined particle/gas cleaning • good for high temperature and humidity
	Gravity settling chambers	• simplest and oldest of designs now rarely used • sticky or highly viscous particles/drops can be collected
	Baffle chambers	• simple design which is more compact than settling chamber now rarely used

Disadvantages	Final form
• low efficiency < 10 µm • problems with sticky materials	Dry powder
• high temperatures needs special bags • potential fire/explosion risk • high maintenance • problems with sticky/hygroscopic particles and condensation	Dry powder
• high capital costs • sensitive to operating variations • unsuitable for very high/low resistivity particles • large space needed • potential explosion hazard • high voltage hazard • sophisticated maintenance	Dry powder
• may create scrubbing liquid disposal problem • potential corrosion problems • high pressure drop • high maintenance • visible plume	Dry powder
• relatively large volume • only really effective for particles > 100 µm	Dry powder
• relatively large volume • only really effective for particles > 100 µm	Dry powder

TABLE 6.4 (continued)
Gaseous treatment technologies

Waste component	Technique	Advantages
Droplets	Impingement and waveplate demisters	• wetted wall and zig zag design • bets for large drops > 30 μm
	Wet cyclones	• similiar to ordinary cyclones can handle solid particles and high liquid loads
	Wet scrubbers	• a wide range of plate, venturi, spray and other devices
Gas purification	Absorption	• packed or plate columns • solvent selection important (water most common) • low pressure drop • plastic construction for corrosive duties • compact • cheap • limited combined particle/gas cleaning
		• spray towers offer low pressure drop and handle particulates well
		Other options include: • sparged tanks • stirred tanks • venturi scrubbers • falling film units • combined hypochlorite absorption with catalytic oxidation system developed for odours and VOCs
	Adsorption	• packed fluidized bed columns • adsorbent selection important (activated carbon, zeolites, alumina, silica gel most common) • flexible to operating conditions • good for recovery duties and trace contaminants

Disadvantages	Final form
• re-entrainment and flooding must be considered	Drained liquid
• special design consideration over re-entrainment and flooding	Drained liquid
• re-entrainment and flooding must be considered	Drained liquid
• may create liquid disposal problem • need for stripping stage if solvent is recycled • prone to plugging from particles • high maintenance	Contaminated liquid plus possibly wet solids
• short residence time device	Contaminated liquid plus possibly wet solids
• regeneration of adsorbent • degradation of adsorbent • high capital cost • need to remove particulates upstream and cool to operating temperature	Contaminated adsorbent

TABLE 6.4 (continued)
Gaseous treatment technologies

Waste component	Technique	Advantages
Gas purification (continued)	Adsorption (continued)	• dry scrubbing by dispersion of fine adsorbent particles into gas stream offers high surface area • can be retrofitted to existing gas cleaning devices
	Condensation	• good for pretreatment recovery of condensable component(s) especially for reuse
	Combustion/ incineration (see Section 6.6, page 135)	• flares (should be used mainly intermittently/ excess hydrocarbon gases) • thermal incinerators (mainly for low concentration organics) • catalytic incinerators/converters (lower temperature operation) • simple • potential energy generation/recovery • effective destruction of organics
	Spray drying	• major application in flue gas desulphurization (FGD)
	Wet electrostatic precipitators	• use absorbent sprays with claimed enhanced performance from electric field • useful for combined removal of fine particulates, mists and gaseous contaminants

shows schematically the main features of each technique. The information required for selection of particulate collection equipment is listed in Table 6.5, page 117. Clearly, any gaps in data will have to be obtained or assumptions justified. A short-list of possible options can be generated in broad terms from Tables 6.6, 6.7 and 6.8, pages 118–123.

The choice of options for removing non-methane VOCs from gas streams depends upon the initial concentration, the gas flowrate, whether only one or a number of species are present and, most importantly, on whether there is value in recovering the VOCs for reuse.

The unit operations for VOC control may be found in Table 6.4. Thermal

Disadvantages	Final form
• need for particulate adsorbent collection • prone to attrition of adsorbent • equipment may be larger than other devices	Contaminated adsorbent
• low efficiency • prone to fouling	Pure liquid contaminant
• potentially high operating costs • explosion hazard • incomplete combustion hazard • poor UK public image due to perceptions over dioxins and air emissions	Gaseous emissions and ash
• large volume chambers • high usage of scrubbing solids	Dry powdered solids
• high costs • electrical hazard	Contaminated liquid absorbent

swing adsorption using either activated carbon or hydrophobic zeolites can be used for one of three purposes:

(1) to concentrate and subsequently to recover VOCs by condensation from the desorbing fluid; if steam is used as the desorbing fluid then preferably the VOCs should not be soluble in water;

(2) as for (1), except that the condensed VOCs can be disposed of by landfill or liquid incineration;

(3) to concentrate the VOCs in an air stream which is to be fed into an incinerator, either thermal or catalytic; preconcentration in this way reduces the size, and the support fuel requirement, of the incinerator.

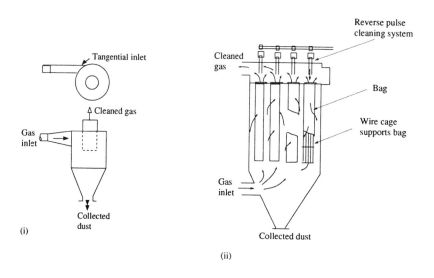

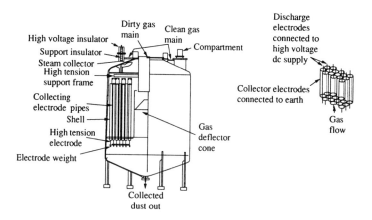

Figure 6.2 Gas treatment equipment: (i) Cyclone*; (ii) Bag filter*; (iii) Electrostatic precipitator*. (* Courtesy of SPS, AEA Technology.)

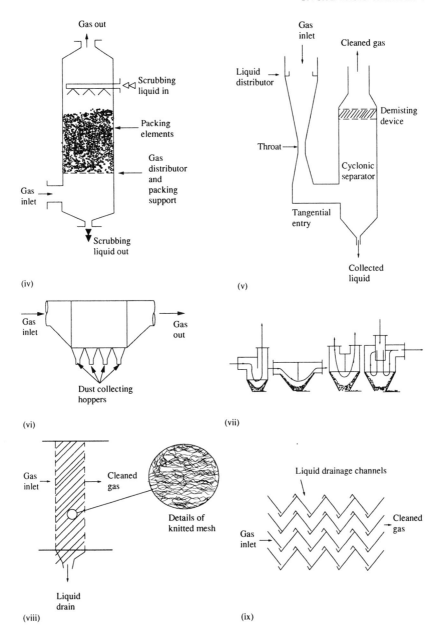

Figure 6.2 Gas treatment equipment: (iv) Wet scrubber – packed tower*;
(v) Wet scrubber – venturi; (vi) Gravity settling chamber*; (vii) Baffle chambers*;
(viii) Demister pad; (ix) Waveplate demisters*. (* Courtesy of SPS, AEA Technology.)

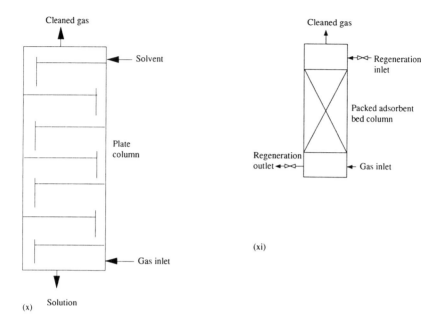
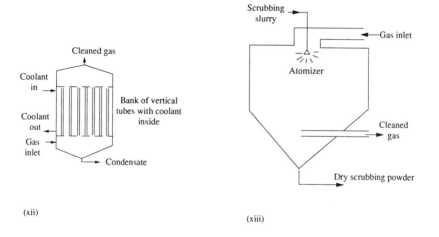

Figure 6.2 Gas treatment equipment: (x) Absorption; (xi) Adsorption; (xii) Condensation; (xiii) Spray drying.

TABLE 6.5
Information required for the selection of particulate collection equipment[†]

Technical requirements	• particle phase (solid, liquid or mixed) • analytical cut diameter required • gas temperature at inlet to collector • corrosiveness of gas • dust inlet burden • erosiveness of the particles • stickiness of the particles • particle density
Process requirements	• plant turndown • plant turnup • constant pressure operation • high pressure operation • continuous cleaning • dry product • simultaneous gas removal (gas purification) • fire and explosion resistance • minimum capital cost of plant • minimum operating cost of plant • minimum complexity • maximum reliability • minimum maintenance

[†] Courtesy of SPS, AEA Technology

Vacuum swing adsorption with either activated carbon or hydrophobic zeolites can be used in combination with condensation to recover VOCs in liquid form, either for reuse or for disposal by landfill or incineration. The advantages of hydrophobic zeolites over carbon are that they work well in humid air streams, can be regenerated in oxidizing atmospheres and do not ignite when certain organic compounds — for example, ketones — are present.

Absorption can be used to transfer VOCs into the aqueous phase either for biological treatment or for chemical oxidation, in which case the process may be enhanced by the use of catalysts, UV light, hydrogen peroxide and soon.

Incineration can be used for the virtually complete destruction of VOCs, but if halogenated species are present then more severe incineration conditions and gas clean-up are required. Recuperative and regenerative forms of incinerator are available. Catalytic devices may be used when lower temperatures are desirable for economic or technical reasons.

TABLE 6.6
Particulate collection equipment[†]

Equipment	Particle phase			Aerodynamic analytical cut diameter required, microns				Gas temperature at inlet to collector, °C	
	Solid	Liquid	Mixed	> 10	< 10	< 3	< 1	< 400	250–400
Demister:									
• inertial	No	Yes	Care	Yes	Care	Care	No	Yes	Yes
• candle filter	No	Yes	Avoid	Avoid*	Yes	Yes	Yes	No	No
Dry inertial collector	Yes	No	No	Yes	Care	No	No	Yes	Yes
Cyclone	Yes	Yes	Care	Yes	Care	No	No	Yes	Yes
Wet deduster:									
• wetted surface	Yes	Yes	Yes	Yes	Yes	Care	No	Care	Yes
• distributed liquid	Yes	Yes	Yes	Avoid	Yes	Yes	Care	Care	Yes
Fibrous filter	Yes	No	No	Avoid	Yes	Yes	Yes	No	No
Fabric filter	Yes	No	No	Avoid	Yes	Yes	Yes	No	Care
Other filter:									
• porous	Yes	No	No	Avoid*	Yes	Care	Care	Care	Care
• granular	Yes	Care	Care	Avoid	Yes	Yes	Care	Yes	Yes
Electrostatic precipitator:									
• wet	Yes	Yes	Yes	Avoid	Care	Yes	Yes	No	Care
• dry	Yes	No	No	Avoid	Care	Yes	Yes	Care	Yes

Yes — plant is generally suitable
Care — plant can be made suitable by special care in its design and operation
Avoid — plant could present problems and other alternatives are normally sought
* usually avoided for economic reasons

[†] Courtesy of SPS, AEA Technology

Gas temperature at inlet to collector			Inlet burden, cm m^{-3}			Particle properties		
Above dewpoint	Near dewpoint	Corrosive gas	> 500	< 500 > 0.2	< 0.2	Erosive	Sticky	Low density or fluffy
Yes	Yes	Yes	Care	Yes	Yes	Care	Avoid	—
Yes	Yes	Care	No	Care	Yes	—	Avoid	—
Yes	Avoid	Yes	Yes	Yes	Yes	Care	Avoid	Avoid
Yes	Avoid	Yes	Yes	Yes	Yes	Care	Care	Care
Yes	Yes	Yes	Avoid	Yes	Yes	Yes	Yes	Yes
Yes	Yes	Yes	Care	Yes	Yes	Care	Yes	Yes
Yes	Avoid	Care	No	Avoid	Yes	Yes	Avoid	Yes
Yes	Avoid	Care	Care	Yes	Care	Care	Avoid	Care
Yes	Avoid	Care	Avoid	Yes	Yes	Care	Care	Yes
Yes	Avoid	Yes	Avoid	Care	Yes	Yes	Care	Yes
Yes	Yes	Care	Care	Yes	Yes	Yes	Yes	Yes
Yes	Avoid	Care	Care	Yes	Yes	Yes	Avoid	Care

TABLE 6.7
Process screening of particulate collection equipment[†]

Equipment	Process requirements					
	Up to 25% turndown required	Up to 25% turnup required	Constant pressure drop across plant required	High pressure operation	Continuous on-line cleaning	Dry product required
Demister						
• inertial	Avoid	No	Yes	Yes	Yes	—
• candle filter	Yes	Care	Yes	Yes	Yes	—
Dry inertial collector	Avoid	No	Yes	Yes	Yes	Yes
Cyclone	Yes	Yes	Yes	Yes	Yes	Yes
Wet deduster						
• wetted surface	Care	Care	Yes	Yes	Yes	No
• distributed liquid	Care	Care	Yes	Yes	Yes	No
Fibrous filter (CHEPA filter)	Care (Yes)	Care	Care	Yes	Care (No)	No
Fabric filter	Yes	Avoid	Care	Care	Yes	Yes
Other filter						
• porous	Yes	Care	Care	Yes	Avoid*	Care
• granular	Yes	Care	Care	Yes	Care	Yes
Electrostatic precipitator						
• wet	Yes	Avoid	Yes	Care	Yes	No
• dry	Yes	Avoid	Yes	Care	Yes	Yes

Yes — plant is generally suitable
Care — plant can be made suitable by special care in its design and operation
Avoid — plant could present problems and other alternatives are normally sought
* — for ceramic filter 'Care'

[†] Courtesy of SPS, AEA Technology

ON-SITE WASTE TREATMENT

			Process requirements				
Simultaneous gas removal	Fire or explosion risk	Minimum capital costs	Minimum operating cost	Low technical complexity	High required reliability	Low maintenance required	
No	Care	Yes	Yes	Yes	Yes	Yes
Avoid	Care	Care	Care	Yes	Yes	Yes
No	Care	Yes	Yes	Yes	Yes	Yes
No	Care	Yes	Yes	Yes	Yes	Yes
Yes	Yes	Yes	Yes	Yes	Care	Care
Yes	Yes	Yes	Avoid	Yes	Care	Care
No	Care	Yes	Yes	Yes	Yes	Yes
No	Care	Care	Care	Care	Care	Care
No	Care	Yes*	Yes*	Yes*	Care	No
Care	Care	Avoid	Care	Care	Yes	Care
Avoid	Care	Avoid	Yes	Avoid	Care	Avoid
No	Avoid	Avoid	Yes	Avoid	Care	Avoid

TABLE 6.8
Sensitivity of particulate collection equipment types to variations in problem statement parameters[†]

	Operable flow range,[a] % of design	Temperature	Dew point	Pressure
Demisters	80/120	N	N	N
Dry inertial collectors	80/120	N	10°C margin	N
Cyclones	25/125	N[b]	10°C margin	N
Wet dedusters	80/120	N[b]	N	N
Fibrous filters	80/120	S[c]	10°C margin	N
Fabric filters	0/110	S[c]	10°C margin	N
Other filters:				
• porous	0/110	N[b]	10°C margin	N
• granular	0/110	N[b]	10°C margin	N
Electrostatic precipitators:				
• wet	30/110	± 30°C	N	N
• dry	30/110	± 30°C	10°C margin	N

N — Not sensitive, approximation would generally suffice
S — Sensitive, precise determination preferable
[a] determination of actual flow requires accurate measurements of temperature and pressure
[b] limited by materials of construction
[c] sensitivity only important close to limitation of media
Note — Numbers refer to maximum allowable variation from design specifications

[†] Courtesy of SPS, AEA Technology

Gas composition	Phase/ stickiness	Dust concentration	Analytical cut diameter $^d AC_a$	Dust density	Dust composition
S^b	S	10%	10%	15%	N^b
S^b	S	10%	10%	15%	N^b
S^b	S	10%	10%	15%	N^b
S^b	N	20%	10%	15%	S
S^{bc}	S	10%	S	N	S^c
S^{bc}	S	20%	N	N	S^c
S^b	S	20%	S	N	S^c
S^b	S	20%	S	N	N^c
S^b	N	20%	N	N	N
S^b	S	20%	N	N	S

6.4 LIQUID TREATMENT TECHNOLOGIES

In general, biological treatment systems are employed for most liquid treatment applications. Some form of pretreatment may be required — for example, to control temperature or remove metal species. If biological systems cannot cope because the waste stream is toxic or difficult to degrade, then a combination of physical and chemical operations need to be considered — for example, a two-stage system could use precipitation plus flocculation and gravity settling to clarify the waste stream and then adsorption to remove any trace organics.

In many cases, the liquid treatment system produces a sludge which then becomes the principal waste stream. The organic sludge produced in treatment systems may consist of a residue of dead and decaying biological matter. These sludges often require further treatment before they can be disposed of by spreading on land, landfill, sea disposal (which is being phased out by 1998) or incineration. The sludges may require volume reduction and stabilization. Concentration by filtration, centrifugation, gravity thickening and drying are the normal steps. Digestion (anaerobic or aerobic) and conditioning (chemical or thermal) can aid stabilization and increase the solids content. The disposal of sludge is becoming more difficult and increasing degrees of treatment will have to be implemented in the future. Of particular concern are sludges containing heavy metals — that is, cadmium, mercury, etc — which need special treatment, such as electrolysis, to reduce concentrations to acceptable limits.

In future, there will be greatly increased on-site treatment of effluents rather than discharge to sewer or water body, as costs rise and benefits of water reuse are realized. There is also a reduced risk of prosecution as well as a move to space-saving (intensified) technologies. Interest is growing in biotreatment, microwave treatment and longer term supercritical fluid methods for removing contaminants. For aqueous streams which contain organic chemical concentrations too high for biological treatment but too low for effective use of incineration, wet air oxidation may be a practical solution. Widely used in the USA for sludge destruction, it has found a small niche market in the UK. Problems are associated with the cost of the equipment, which must withstand elevated pressures and temperatures, and with corrosion and materials of construction.

The wide range of devices available for removal of particulates and contaminants from liquid streams is related to molecular and particle size in Figure 6.3. Table 6.9, pages 126–135, lists the treatment options and briefly covers their advantages and disadvantages. The various treatment technologies are classified into biological, chemical and physical systems. Within each group, an attempt has been made to place those techniques which are most suitable for processing liquids first, progressing through to concentrated liquids and finally liquid sludges. Figure 6.4, pages 136–144, shows schematically the main

ON-SITE WASTE TREATMENT

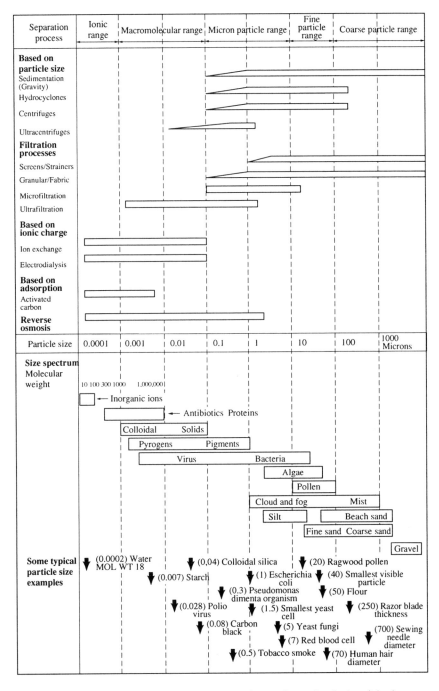

Figure 6.3 Classification of liquid treatment equipment by molecular/particle size.

125

TABLE 6.9
Liquid treatment technologies

Classification	Technique	Advantages
Biological	Lagoons	• basins 2–5 m deep • aerated mechanically or by blowing • simple design
	Bio-filters (trickling filters)	• large surface area • plastic packings • promote aerobic bacteria • irrigated by fixed or rotary distributors • low energy system • flexible • widely used for 'roughing' duty
	Rotating bio-contactors	• mechanically rotated devices with discs/plastic packing • compact • good for low/medium strength wastes
	Anaerobic digesters	• contact with bacteria excluding oxygen • good for high strength wastes — for example, food, agriculture, brewing and distilling
	Activated sludge equipment	• vigorous aeration of floc followed by sedimentation • yields carbon dioxide and water • widely used • good at reducing BOD and COD from medium strength organic wastes • yields high quality treated water
	Composting	• decomposition and stabilization of high organic content waste on an organic support — for example, wood shavings • environmentally tolerant
Chemical	Ion-exchange	• cation exchange • resins effective • removal of metals to < 1µg/l from dilute solution • highly selective

Disadvantages	Final form
• very large land area • malodours	Sludge
• prone to biocides • packing needs to be kept wet by recycling at low throughput • needs downstream clarifier • malodours • activated sludge yields better quality	Sludge from clarifier
• mechanical problems	Sludge
• may need further polishing • bacteria can be poisoned • strong odours	Bio-gas (fuel?) high strength sludge
• very large land area • malodours	
• may need prior dewatering • potential handling problems • labour intensive	Compost
• relatively expensive • regeneration chemicals needed • risk of poisoning	Concentrated solution

TABLE 6.9 (continued)
Liquid treatment technologies

Classification	Technique	Advantages
Chemical (continued)	Electrolysers	• similiar to electroplating devices • waste must be good electrolyte • recovery of metals • destruction of cyanides
	Neutralizers	• dosing with acid or alkali to adjust pH • can use waste acids/alkalis
	Wet air oxidation	• effective on phenols, insecticides and pesticides • pH lowered to typically 4 by adding acid then pressure increased to around 50 bar before passing into the first of two identical oxidation reactors • heat recovery • gas/liquid separator yields mainly nitrogen and carbon dioxide to atmosphere • separated liquid is treated to remove any metal contaminants and goes to a final polishing plant before discharge to sewer
	Reduction equipment	• dosing with reducing agents — for example, ferrous sulphate, sulphur dioxide or sodium metabisulphite • needs acidic conditions • useful pretreatment — heavy metals removal
	Oxidation equipment/ chlorinolysers /disinfectors	• dosing with chlorine (often hypochlorite), hydrogen peroxide and ozone • UV light often added • useful partial treatment — for example, cyanides and chlorinated pesticides
	Precipitators	• render insoluble component for recovery or disposal by addition of chemicals such as lime, ferric sulphate and chloride, ferric sulphate and aluminium sulphate • useful partial treatment • can bring down other suspended and colloidal matter

Disadvantages	Final form
• expensive to operate • electrodes may need replacing	Recovers metals, by-products, oils/fats/solids for disposal
• needs careful control • cost of chemicals • toxic gases may be released — for example, when the waste contains sulphides or cyanides	Fluid
• high costs with complex plant to operate and maintain	Fluid
• pH sensitive • may need large volumes of acid	Chemical sludge
• cost of chemicals • safety hazards	Chemical sludge
• chemistry needs to be considered • cost of chemicals • additional flocculation stage needed • dosage rates critical	Chemical sludge

TABLE 6.9 (continued)
Liquid treatment technologies

Classification	Technique	Advantages
Chemical (continued)	Flocculators	• coagulation of small particles aided by polyelectrolytes • useful after precipitation and for minerals/clay
Physical	Reverse osmosis	• semi-permeable membrane used to separate water from salt solutions (desalination) • useful for metals removal from plating wastes
	Electro-dialysis	• membrane + electric field separation of ions in dilute solutions — for example, desalination, plating effluents
	Air strippers	• often conventional cooling towers • use for carbon dioxide, ammonia, hydrogen sulphide, chlorinated hydrocarbons and VOCs • detergent foams • low capital and operating costs • simple
	Steam stripping	• higher temperatures than air stripping • organics recovered as separate liquid phase • very effective for VOCs
	Solvent extraction	• use of selective solvent for recovery of valuable components or phenols
	Adsorption	• often into activated carbon beds • useful for non biodegradable solubles such as colouring and phenols
	Oil/water separators	• insoluble oils/greases can be separated by gravity skimmers
	Ultrafiltration	• membrane process for molecular weights > 500 (macromolecules, colloidal particles and emulsions) • less expensive than reverse osmosis
	Equalization	• pretreatment to ensure uniform flow/ concentration • holding and mixing tanks

Disadvantages	Final form
• high cost of polyelectrolytes • additional solids separation stage needed	Chemical sludge
• high capital cost • prone to membrane fouling/degradation	High concentration salts
• relatively expensive • prone to membrane fouling/degradation	Concentrated solution
• may need pH adjustment to 'free' soluble gases	Contaminated air stream may need further treatment — for example, combustion or dilution
• higher capital and operating costs than air stripping • more complex design	Chemical sludge
• high capital and solvent costs	Raffinate high in solids plus residual solvent
• need to regenerate beds	Contaminated carbon
• problems of emulsions and foaming	Fluid/emulsion
• prone to membrane fouling/degradation	Sludge/gel
• large volume	Untreated waste stream for further processing

TABLE 6.9 (continued)
Liquid treatment technologies

Classification	Technique	Advantages
Physical (continued)	Evaporation	• heating to concentrate solids • useful if solids can be recovered
	Flotation	• air bubbles dispersed into agitated suspension with hydrophobic particles collected at surface (widely used in minerals processing)
	Filtration	• simple to operate • can have low capital and running costs • wide range of vacuum, rotary, pressure and belt devices
	Centrifuges	• continuous conical bowl • enclosed • high throughput • high concentration • not prone to clogging • polymer dosing may allow up to 45% solids
	Deep-well injection	• pumping into permeable rock formations — for example, limestone or dolomite — or caverns • similiar to oil or gas well technology • used for difficult wastes
	Sedimentation	• simple gravity settling of treated sludges prior to disposal
	Hydro-cyclones	• centrifugal separators • simple • good for coarser sludges
	Dryers	• thermal separation of liquids from solids • large choice of equipment • significant volume reduction • useful prior to incineration • can yield fertilizer
	Screening systems	• simple fixed, vibratory and rotary devices widely used for removal of coarse solids

Disadvantages	Final form
• high capital and energy costs	Wet solids
• may require wetting, frothing and deflotation agents	Sludge
• filter media may 'blind' • precoat may be needed • high energy costs for vacuum systems • back washing may add to costs	Filter cake
• high energy and maintenance costs • noisy • needs prior screening and de-gritting	Sludge cake and concentrate may need further treatment
• need for buffers • pH adjustment	
• high capital cost and space • labour intensive • sensitive to distribution and flow patterns	Thickened sludges and supernatant
• need careful design • need consistent flowrate	Sludge and concentrate may need further treatment
• high energy costs if cheap waste fuel is not used • risk of air pollution in convective drying systems	Dry solids
• screen size critical • prone to wear/failure	Wet coarse solids may need further treatment

MANAGEMENT OF PROCESS INDUSTRY WASTE

TABLE 6.9 (continued)
Liquid treatment technologies

Classification	Technique	Advantages
Physical (continued)	Combustion /incineration (see Section 6.6).	• complete destruction of high concentration organics • particularly toxic materials

features of each technique. The information required for selection of liquid treatment equipment is listed in Table 6.10, page 145. Clearly, any gaps in data will have to be filled or assumptions justified. A short-list of possible options can be generated in broad terms from Figure 6.5, page 146–147.

6.5   SOLIDS TREATMENT TECHNOLOGIES

Depending on the physical characteristics of the solid waste, it may require a primary treatment step in order to homogenize it as far as possible. It is also likely that a solid waste treatment system will produce a residue of material which may require disposal by landfill.

The main alternative to landfill is incineration which is an effective method of reducing the waste volume by a factor usually greater than 90% with a considerable reduction in weight. Although landfill can be managed on large sites, it is generally regarded as an off-site disposal route and is discussed in detail in Section 7.5, page 162. Incinerators are operated both directly by industrial concerns on site and by merchant (commercial) waste disposal companies. Most people are familiar with the development of municipal waste incinerators. Section 6.6, page 135, discusses incineration in detail.

Recent developments which may impact on incineration are the use of molten salts and metals. A US company has pioneered the molten metal technology and several commercial plants are nearing completion[3]. It is claimed to offer extremely low environmental impact with the opportunity to convert wastes into usable materials such as synthesis gas, metal alloys and ceramics. The process is also claimed to be much cheaper than incineration. A new plant for Hoechst Celanese in Texas will treat wastewater sludges, solvents and other wastes at up to 22,000 t yr^{-1}.

A special form of solid waste which should be considered is scrap

ON-SITE WASTE TREATMENT

Disadvantages	Final form
• non-combustible components can cause problems • incomplete combustion is hazardous • may need additional fuel • metallic components can cause emission problems • need for expensive gas cleaning systems	Combustion products and ash

equipment and plant. This can be thought of in exactly the same manner as any other form of solid waste. Increasingly as product life cycles shorten, equipment and plants must become more flexible and reusable. However, care and planning should be given at the earliest stages of equipment/plant design and procurement on its potential for reuse and ultimate disposal. This may involve decommissioning and an IChemE guide on this topic is available[4].

Table 6.11, page 148, lists the treatment options and briefly covers the advantages and disadvantages. The information required for selection of solids treatment equipment is listed in Table 6.12, page 150. Clearly, any gaps in data will have to be filled or assumptions justified. A short-list of possible options can be generated in broad terms from Figure 6.6, page 155.

6.6   INCINERATION

If the process plant has waste arising from a number of process operations in several different phase forms, the most appropriate technology appears to be incineration, provided there is a clear understanding of the composition of any waste stream or mixture of streams. Unfortunately, incineration has a poor public image, exacerbated by concerns over possible problems with dioxins in some applications (those where chlorine is present). However, in the UK it has received the backing of the Royal Commission for Environmental Pollution[5]. A recent review of all aspects of waste incineration has been produced by the Royal Society of Chemistry[6]. The main European Council Directive on the incineration of hazardous waste 94/67/EC[7] (see Section 3.11.6, page 52) provides measures to prevent and reduce the adverse environmental effects. Also, the EU is supporting a project called Incipro to develop clean incinerator technology which, if successful, should improve the environmental acceptability of the process.

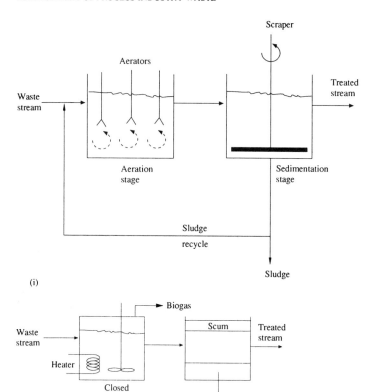
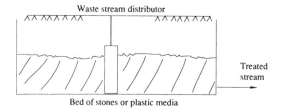

Figure 6.4 Liquid treatment equipment: (i) Activated sludge process; (ii) Anaerobic digester; (iii) Bio/filter/trickling filter.

ON-SITE WASTE TREATMENT

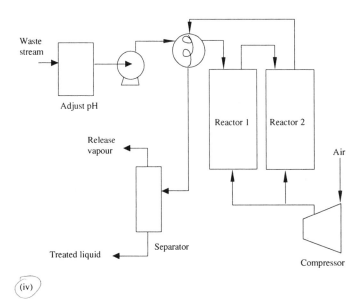

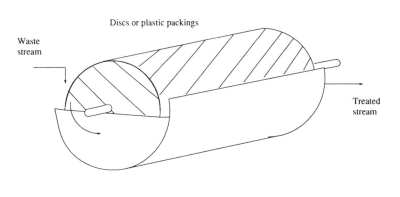

Figure 6.4   Liquid treatment equipment: (iv) Wet air oxidation; (v) Rotating biological contactor.

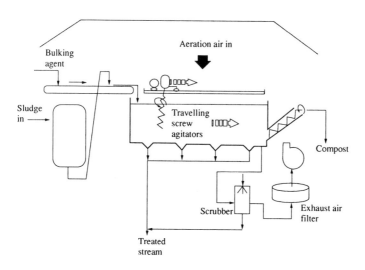

(vi)

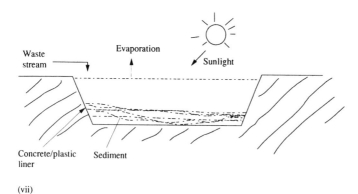

(vii)

Figure 6.4 Liquid treatment equipment: (vi) Composter (courtesy of the Effluent Processing Club (EPC), AEA Technology); (vii) Lagoon.

ON-SITE WASTE TREATMENT

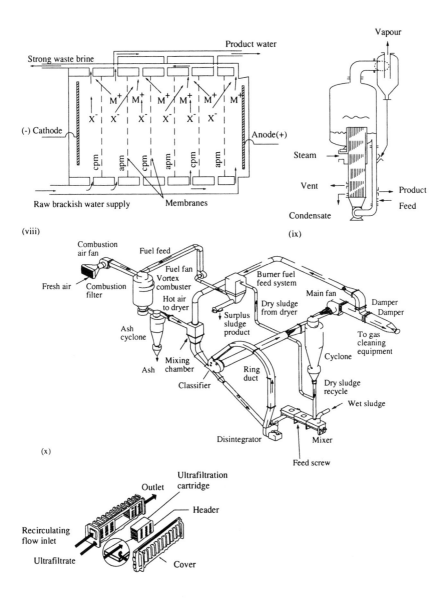

Figure 6.4 Liquid treatment equipment: (viii) Electro-dialysis (Scott & Smith, 1981, *Dictionary of Waste and Water Treatment*, reproduced by permission of Butterworth-Heinemann Ltd, Oxford, UK); (ix) Evaporator (Perry, R. and Green, D., 1984, *Perry's Chemical Engineers' Handbook*, 6th edition, reproduced by permission of The McGraw-Hill Companies); (x) Dryer (courtesy of Barr-Rosin Ltd); (xi) Ultrafilter.

MANAGEMENT OF PROCESS INDUSTRY WASTE

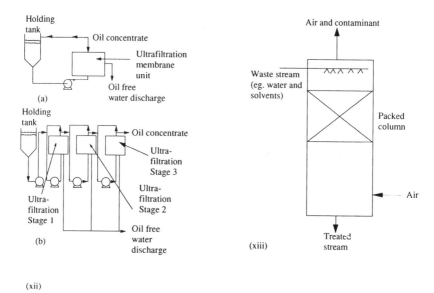

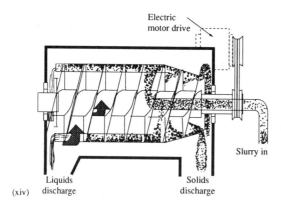

Figure 6.4 Liquid treatment equipment: (xii) (a) Batch ultrafiltration, (b) Continuous ultrafiltration; (xiii) Air stripper; (xiv) Centrifuge (Scott & Smith, 1981, *Dictionary of Waste and Water Treatment*, reproduced by permission of Butterworth-Heinemann Ltd, Oxford, UK).

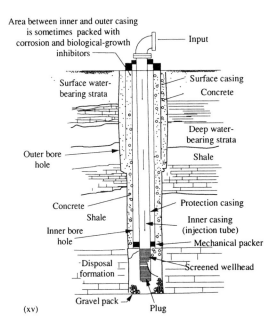
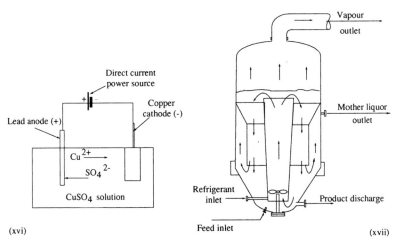

Figure 6.4 Liquid treatment equipment: (xv) Deep well injection*; (xvi) Electrolyser[†]; (xvii) Precipitator*. (* Perry, R. and Green, D., 1984, *Perry's Chemical Engineers' Handbook*, 6th edition, reproduced by permission of The McGraw-Hill Companies). ([†] Scott & Smith, 1981, *Dictionary of Waste and Water Treatment*, reproduced by permission of Butterworth-Heinemann Ltd, Oxford, UK).

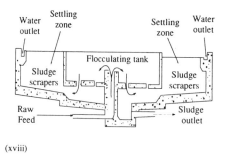

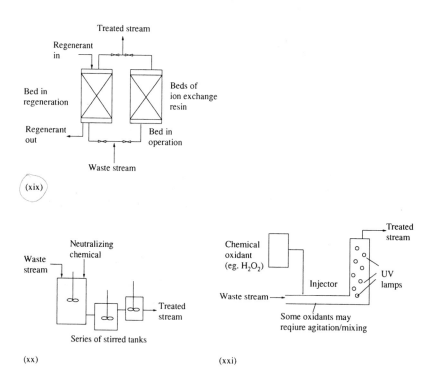

Figure 6.4 Liquid treatment equipment: (xviii) Flocculator (Scott & Smith, 1981, *Dictionary of Waste and Water Treatment*, reproduced by permission of Butterworth-Heinemann Ltd, Oxford, UK); (xix) Ion exchange system; (xx) Neutralizer; (xxi) Oxidation system.

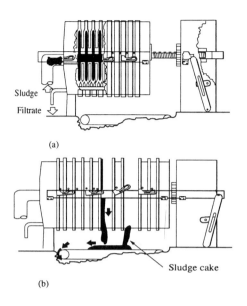
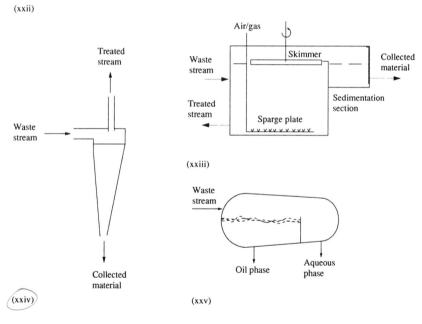

Figure 6.4 Liquid treatment equipment: (xxii) Filter press: (a) Injection of sludge between plates before the sludge is de-watered; (b) sludge cake discharge (Scott & Smith, 1981, *Dictionary of Waste and Water Treatment*, reproduced by permission of Butterworth-Heinemann Ltd, Oxford, UK); (xxiii) Flotation; (xxiv) Hydrocyclone; (xxv) Oil/water separator.

MANAGEMENT OF PROCESS INDUSTRY WASTE

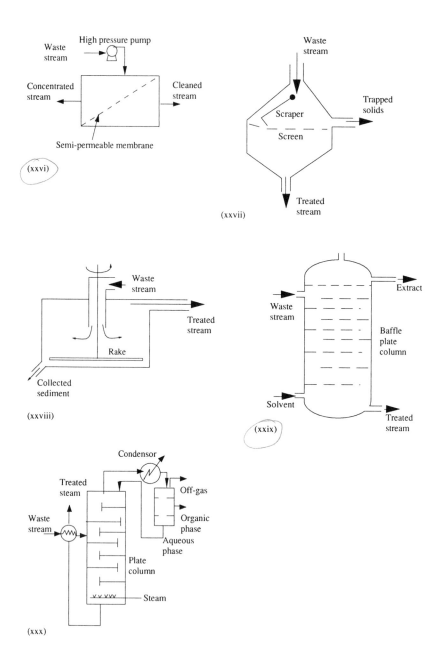

Figure 6.4 Liquid treatment equipment: (xxvi) Reverse osmosis; (xxvii) Screening system; (xxviii) Sedimenter/clarifier; (xxix) Solvent extraction; (xxx) Steam stripper.

TABLE 6.10
Liquid waste stream data specification[†]

Question	Analyses required
1a Does the manufacturing process involve inorganic raw materials, by-products or end products?	• total metals • alkalinity • COD • total dissolved solids • other specific contaminants
1b Does the manufacturing process involve organic raw materials, by-products or end products?	• TOC • BOD (COD optional) • oil and grease or TPH • other specific contaminants
2 Does the process generate waste streams that are acidic or caustic?	• pH • buffering capacity
3 Does the process generate high temperature waste streams?	• temperature
4 Does the waste stream contain entrained solids?	• total solids • total suspended solids • total dissolved solids • turbidity
5 Does the waste stream contain nitrogen compounds?	• NH_3 • NO_3 • total Kjeldahl nitrogen
6 Does the waste stream contain cyanide compounds?	• total cyanide • reactive cyanide
7 Does the waste stream contain sulphur compounds?	• sulphides • sulphates • sulphites
8 Does the waste stream contain phosphorus compounds?	• phosphates
9 Does the waste stream contain surfactants or have excessive foaming?	• surfactants
10 Does the waste stream contain any toxic compounds?	• total toxic organics • toxic metals

[†] Belhateche, D.H., 1995, Choose appropriate wastewater treatment technologies, *Chem Eng Prog*, 91 (8): 32–51. Reproduced with permission of the American Institute of Chemical Engineers. Copyright © 1995 AIChE. All rights reserved.

MANAGEMENT OF PROCESS INDUSTRY WASTE

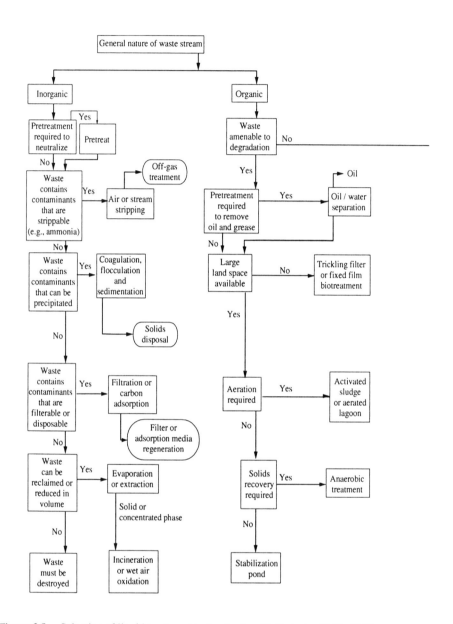

Figure 6.5   Selection of liquid treatment technologies. (Belhateche, D.H., 1995, Choose appropriate wastewater treatment technologies, *Chemical Engineering Progress*, 91 (8): 21–51. Reproduced with permission of the American Institute of Chemical Engineering. Copyright ©1995 AIChE. All rights reserved.

ON-SITE WASTE TREATMENT

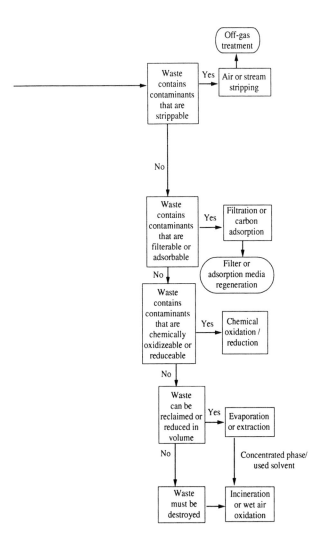

147

TABLE 6.11
Solids treatment technologies

Classification	Technique	Advantages
Mechanical	Compactors	• crusher to reduce volume • typical final density 1100 kg m^{-3}
	Encapsulation/ stabilization	• hazardous waste is transformed into a durable solid product with very low leaching characteristics fixed by adding solidifying agent to the waste — the agent can be either silicate (cement) based, glass or an organic polymer
Thermal	Incineration (see Section 6.6, page 135)	• complete combustion destroys organic wastes • offers energy from waste
	Pyrolysis/ gasification	• decomposition by heating excluding oxygen • scope to recover useful materials such as carbon and fuels
	Calcination	• high temperatures of about 1000°C • dry powder from slurries • sludges and tars — for example, refinery sludge containing hydrocarbons
	Composters	• decomposition and stabilization of high organic content waste on an organic support — for example, wood shavings • environmentally tolerant

Interest in the combustion of waste as an additional source of energy has been investigated by ETSU (Energy Technology Support Unit), based at AEA Technology, Harwell. ETSU has produced an extensive series of reports on the topic. Additionally, life cycle analysis of waste paper has concluded that it may be environmentally better to burn rather than recycle the paper[8].

The process of incineration is essentially one of thermal oxidation. Depending on the amount of oxygen available, it can be described as starved air (pyrolysis/gasification) or direct combustion (incineration) operation. The principles of combustion are well understood for the fuel burners but the reaction schemes can be very complex — for example, some 40 reactions have been

Disadvantages	Final form
• cannot be used for incompressible wastes	Compacted solids
• need to find compatible solidifying agent	Solidified block/drum
• non-combustible components can cause problems • incomplete combustion is hazardous • may need additional fuel • metallic components can cause emission problems • need for expensive gas cleaning systems	Combustion products and ash
• risk that recovered materials may contain unacceptable levels of hazardous components	Synthesis gas, oils and tars, carbon, non combustible residues
• energy intensive unless the waste contains a combustible organic	Calcined solids plus products of combustion
• potential handling problems • labour intensive	Compost

identified in the combustion of soot. The key '3-T' design criteria are Temperature, residence Time and Turbulence which are interdependent. In most waste incinerators, high temperatures and reasonably long residence times with high turbulence provide a large design margin to ensure that all combustible components are decomposed and converted to carbon dioxide, water, sulphur dioxide, hydrogen chloride, chlorine, phosphorus pentoxide and nitrogen. Some of these products of combustion are corrosive, so the downstream gas cleaning system and the stack must be protected by suitable design to keep temperatures well above acid dewpoints and by the selection of appropriate materials of construction.

TABLE 6.12
Solid waste stream data specification

Question		Analyses required
1a	Waste of inorganic origin?	• total metals • alkalinity • other specific contaminants
1b	Waste of organic origin?	• flammability/calorific value • explosive • toxic organics • other specific contaminants
2	Form of waste?	• sludge/paste viscosity • powder particle size • solids bulk density • grindability
3	Is it mixed waste?	• metal items content

Incineration of process waste in the UK is presently dominated by merchant operators. Holmes[9] reports that the UK has some 125,000 tonnes per year capacity from four commercial plants for solids wastes processing, whilst only some 3000 tonnes per year are available for liquid wastes. Additionally, he estimates at least 50 units in chemical companies with their own units including thermal oxidizers, catalytic oxidizers, boilers and incinerators. Operators have dedicated facilities which accept a wide range of waste from the whole of the industrial spectrum and levy a disposal charge related to the nature and quantity of the waste being received.

Due to the wide range of feed materials to the incinerators, operation of the units is tightly controlled and the systems for acceptance, handling, incineration, ash removal and gas clean-up are closely monitored. A feature of merchant incineration is the requirement to bulk up waste and transport it from the production site to the disposal facility. This introduces a general handling hazard.

In-house treatment may include some level of pre-treatment of the waste streams prior to introduction into the incinerator which can also be used to burn fugitive emissions and other gaseous streams with a calorific value. Other major advantages of in-house units include reduced transportation risks and retention of control of the waste by the waste producer. Furthermore, integrated heat recovery reduces the effective disposal costs.

Combustion equipment usually embodies (not exclusively) the following key elements, also shown in Figure 6.7, page 152:
- waste pretreatment;
- waste handling and charging;

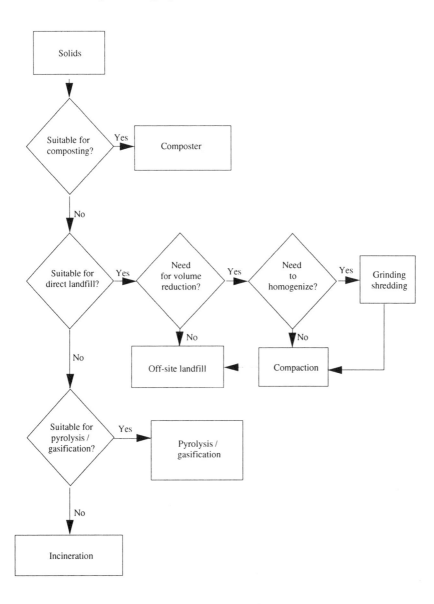

Figure 6.6   Selection of solids treatment technologies.

MANAGEMENT OF PROCESS INDUSTRY WASTE

- primary combustor/afterburner units;
- de-ashing;
- waste heat recovery (optional);
- flue gas clean-up;
- automatic/semi-automatic control and monitoring systems.

Different waste streams may be blended to obtain the desired calorific value which sustains combustion without using supplementary fuel. Such blending is easier to achieve in larger scale merchant incinerators where a large number of waste streams of varying composition are available. Consideration

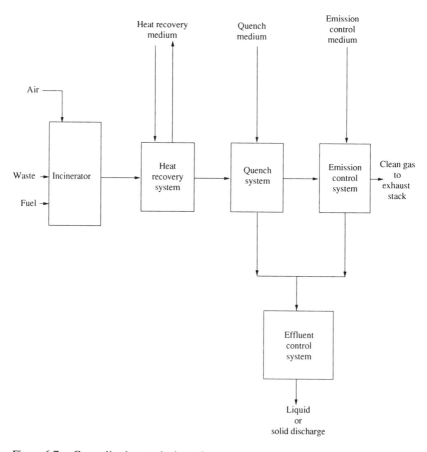

Figure 6.7   Generalized waste incineration system.

152

TABLE 6.13
Incinerator technology

Type	Comment	Operating temperatures,°C
Fixed hearth	• traditional technology stepped or reciprocating grate • mainly solids	800–900
Rotary hearth	• cascade bed • solids and sludges	900–1000
Fluidized bed • bubbling • fast • dual	Small solids and liquids: • traditional — gas velocity 1 ms^{-1} • entrained bed — gas velocity 3 ms^{-1} • combined bed — 2nd generation	800–900
Rotary kiln — slagging	• widely used • mainly solids although some liquids • high temperature • melts inorganics	>1200
Rotary kiln — non slagging	• mainly low melting point solids although some liquids	<900
Down fired	• liquids (with salts / organics and gases)	>1200
Cyclonic	• sludges and liquids — high or low calorific value may contain solids	>1000
Axial	• odours, fumes and gases and some liquids	>1000
Catalytic	• odours and gases	<400

should be given, however, to blending waste streams for smaller on-site incinerators. After blending, some solid waste streams may also need to be reduced in size by shredding. Magnetic separation may be used to separate any metals in the waste stream.

The treated waste is burnt with excess air in the combustion chamber of the incinerator. If the calorific value of the waste is insufficient to sustain combustion then supplementary fuel — for example, fuel oil, natural gas — is needed. The different types of incinerators that can be used for burning specific types of wastes are summarized in Table 6.13. Combustion temperatures range from 800 to 1200°C and sometimes combustion is completed in a secondary chamber downstream of the main incinerators. Secondary combustion is used particularly when solids are being burnt in the primary chamber and/or difficult compounds such as halogenated species are being burnt. There is no clear division between the various incinerator types and potential applications. However, most commercial operators utilize rotary kilns as illustrated in Figure 6.8. Fixed

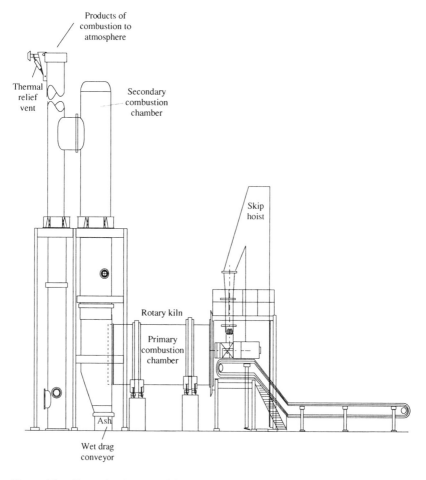

Figure 6.8 Example of a rotary kiln incinerator. (LaGrega, M., 1994, *Hazardous Waste Management*, reproduced by permission of The McGraw-Hill Companies.)

hearth and rotary kilns are generally employed to deal with solids but with suitable design they can be made 'omnivorous' — burning solids, liquids and possibly gases. Liquids and gases, such as vent streams, are usually incinerated in axial or down-fired units, depending on the solids content in the feed streams.

The hot combustion gases from the incinerator are cooled depending on the mode of the heat recovery used. In smaller on-site incinerators, the combustion gases are usually cooled to low temperatures in a waste heat boiler so that steam can be raised for on-site use. In larger merchant incinerators, heat recovery is often not a prime consideration.

It is widely believed that the slow cooling of combustion gases in the temperature range 450 down to 200°C is a major source of the group of chemicals (polychlorinated dibenzofurans and polychlorinated dibenzo-p-dioxins) which collectively are known as 'dioxins'. The Environmental Agency's guidance notes make specific reference to this aspect and set the maximum dioxin concentration in a flue gas (as measured by the NATO/CCMS method) at 1 ng m^{-3} but with a target to reduce this level to 0.1 ng m^{-3}. Of course, dioxins can only be formed if chlorinated species are being fed into the combustion device.

The cooled gases are then scrubbed with water or sodium hydroxide in a scrubber before separating any solid particles in filters or electrostatic precipitators. This cooled, clean gas should meet the appropriate emission consent for discharge to atmosphere through a stack. Liquid streams from the scrubbers are usually sent to conventional biological treatment to clean the contaminated liquid streams before these too can be discharged to the sewer or the aquatic environment.

A recent development in the UK is to utilize existing cement kilns which operate at high temperatures to combust wastes; the practice is well established in other countries such as France. Rhône Poulenc's venture called Scori was Europe's first major waste-to-fuel disposal route via cement kilns (some 30% of the hazardous waste treatment market in France)[10]. The additional capacity could far exceed that offered by the existing commercial incinerator operators, providing considerable competition.

Some of the more recent incineration developments are listed in Table 6.14. These are not yet in wide use but could have a significant impact in the future for particular problems.

TABLE 6.14
Incinerator technology under development

Type	Comment	Operating temperatures,°C
Infrared	• contaminated soils	—
Plasma red	• solids and liquids	>5500
Advanced electric reactor	• contaminated soils	>2300
Rocking kiln	• high temperature • melts inorganics	>1200

REFERENCES IN CHAPTER 6

1. *Croner's Waste Management Guide*, 1991 (Croner Publications, UK).
2. Sittig, M., 1977, *Particulates and Fine Dust Removal* (Noyes Data Corporation, USA).
3. Yates, I.C., 1995, Molten metal gears up for first commercial applications, *Chem Tech Europe*, 2 (5): 10–11.
4. Briggs, M., Buck, S. and Smith, M. (eds), 1997, *Decommissioning, Mothballing and Revamping* (IChemE, Rugby, UK).
5. Royal Commission on Environmental Protection, 1993, *17th Report, Incineration of Waste, Cm2181* (HMSO, UK).
6. Hester, R.E. and Harrison, R.M., 1994, *Waste Incineration and the Environment* (Royal Society of Chemistry, UK).
7. Council directive 94/67/EC of 16 December 1994 on the incineration of hazardous waste, 1994, *Official Journal of the European Communities*, 37(L365): 34–45.
8. Daae, E. and Clift, R., 1994, A life cycle assessment of the implications of paper use and recycling, *Environmental Protection Bulletin*, No 28, 23–25.
9. Holmes, J.R., 1995, *The United Kingdom Waste Management Industry* (Institute of Waste Management, UK).
10. Lorane, A., 1993, Cleaner production methods for industry: the treatment of industrial waste in France, *Waste Age, Int Suppl*, 15–16, 18–20.

7. OFF-SITE WASTE DISPOSAL

7.1 INTRODUCTION

Off-site waste disposal is essentially a downstream activity following on-site treatment as discussed in Chapter 6. Once a waste stream has been treated, it may be possible to discharge it to air, an aquatic environment or land. An overview of Chapter 7 is given in Table 7.1, page 158. This implies that the extent of the treatment is sufficient to reduce the environmental impact to very low levels to meet compliance limits so that no long-term control measures are necessary. If the waste cannot be discharged directly then it must be transported to a commercial waste contractor who will either simply landfill the waste or pre-treat it prior to landfill or incineration. The movement of waste, particularly special waste, may come under the transport of dangerous goods regulations. More stringent controls exist over the movement of waste outside of the producer's country where the waste originates.

The waste disposal market suffers generally from over-capacity and intense competition largely as a result of the success of companies in minimizing and reusing waste and in segregating hazardous wastes. In some areas, new players have entered the market — for example, commercial incinerators have been supplemented by cement kilns burning waste as fuel.

In most cases, landfill is the ultimate resting place for waste which cannot be directly discharged to the environment and where long-term control and management is needed. Although landfill in the UK is often economically the most attractive option, there is a feeling that more encouragement should be given to move to options higher in the waste management hierarchy, such as recycling and recovery. This has led to the planned landfill levy but the likely scale of any effect on the balance between landfill and incineration is unclear.

7.2 DISCHARGES TO AIR

Stacks are the primary route by which gaseous wastes are discharged to air and dispersion from them is a key issue. They simply effect a dilution of any contaminant(s) as the actual mass flowrate of waste is fixed.

Dispersion is governed by permissible limits on ground level concentrations. The design of stacks therefore depends on topography, meteorology and contaminant properties. A number of plume rise calculation methods are

TABLE 7.1
Overview of Chapter 7

7.2, page 157 Discharges to air	• stacks • dispersion • fugitive emissions
7.3, page 159 Discharges to aqueous environment	• sewer • consents • watercourses • sea • incineration at sea
7.4, page 159 Discharges to land	• spraying • spillages/leaks • fire water
7.5, page 162 Landfill	• landfill design and operation • deep wells and mines
7.6, page 164 Liquid/solid waste contractors	• identification of facilities • assessment of their management and operation • assessment and selection of carriers • follow-up of final disposal
7.7, page 168 Transport of waste	• written documentation • labels and markings • regulations • consignment notes
7.8, page 173 Transfrontier shipment of waste	• controls on the import and export of wastes

available and computational fluid dynamic codes are also being used. A number of dedicated dispersion codes are commercially available (see *CEP Software Directory*[1]).

The issue of fugitive emissions and releases during incidents and emergencies should be considered as a component of the gaseous waste discharges. These topics are beyond the scope of this book and the reader is referred to Hesketh and Cross[2] and the AIChE Centre for Waste Reduction Technologies book on VOCs[3].

7.3 DISCHARGES TO THE AQUEOUS ENVIRONMENT

Discharges to sewer are regulated in the UK by the local water plc. A consent to the discharge of trade effluent conditions specifies volume (total and flowrate) and composition under an application from the operator called a trade effluent notice. It can be very convenient but is not always cost effective. Certain types and amounts of industrial effluent are positively welcomed by the water plcs as they yield a chemically balanced sewage which aids treatment.

Discharges to watercourses are regulated by the Environment Agency and are covered in detail in the NRA *Water Quality Series No 17*[4]. These are much more complex and stringent than sewer consents. Certain discharges require public notification and a six-week period for comment followed by up to three months for a decision. Examples of discharge consents are given in Figures 7.1 and 7.2, pages 160 and 161.

Disposal to sea is currently being curtailed and sludge disposal by this route will be phased out by 1998. Remote discharge via long pipelines and reliance on the natural powers of the sea to treat wastes is being constrained by various EC directives on urban wastewater treatment and bathing water.

Discharges to sea also include 'incineration at sea' using specially converted tankers. These often take hazardous wastes such as chlorinated chemicals and disperse hydrochloric acid into the sea. Technically, experts appear to view this as an attractive option compared with incineration on land. Incineration at sea, however, is now frowned upon by public opinion and the EC agreed to stop it in 1994.

7.4 DISCHARGES TO LAND

The waste, usually partially treated, is sprayed over a tract of land upon which disposal is permitted. This method was used historically for most sewage sludge and is still widely used in the UK and in desert areas for crop irrigation/dust suppression. Some potential problems exist because:
- the effluent must not contain any toxic substances;
- the land must not get waterlogged and should be relatively level;
- no nearby potable water resources must be compromised.

Any potential spillages or leaks need to be included when considering waste discharges to land and the potential for groundwater contamination. This is particularly relevant to the transport of liquid wastes on the site and underground waste storage tanks. Another important topic concerns waste generated during a fire, when water used to extinguish the flames may become contaminated. In some sectors — such as pesticide stores — bunds and treatment facilities for fire water are becoming the norm.

NATIONAL RIVERS AUTHORITY

Reference

WATER RESOURCES ACT 1991 — CONSENT TO DISCHARGE

The National Rivers Authority, in pursuance of its powers under the above mentioned Act, HEREBY GIVES CONSENT to the discharge described hereunder subject to the terms and conditions set out below.

Name & Address of Applicant: Any Water Services Co Ltd
                                123 High Street
                                Anytown

Date of Application: 18 June 1992

Date of Consent: 10 October 1992

Description of Discharge: Type: Final Effluent
                                From: Smallish STW
                                To :   Anytown River

Conditions

1. Except with the agreement of the person making the discharge under this consent, no notice shall be served revoking the consent or modifying the conditions before 10 December 1994.

2. The discharge shall consist of treated sewage effluent from an outlet at National Grid Reference XX 8888 8888.

3. The effluent shall derive from domestic sewage from a population of 250 or less and contain no unauthorized trade waste.

4. As far as is reasonably practicable, the works shall be operated so as to prevent:
   a any matter being present in the effluent which will cause the receiving water to be poisonous or injurious to fish or to their spawn, or spawning grounds or food, or otherwise cause damage to the ecology of the receiving waters; and
   b the treated effluent from having any other adverse environmental impact.

5. The Company will operate the works having regard, so far as is relevant, to the guidance set out in the National Water Council's Occasional Technical Paper Number 4, "The Operation and Maintenance of Small Sewage Treatment Works" dated January 1980. In particular, the works shall be maintained properly such that:
   a it remains fully operational except at time of mechanical or electrical breakdown;
   b any such breakdowns shall be attended to promptly and the equipment returned to normal operation as soon as possible; and
   c tanks shall be regularly desludged at sufficient frequency and in such a manner as to prevent problems with septic tanks, rising sludge or excessive carryover of suspended solids.

6. Facilities shall be provided for safe and convenient access to enable Authority's representatives at any time to take samples, carry out flow measurements and inspection to ensure that the conditions of this consent are complied with.

NRA Regional Office   ....
Address   NRA Authorized Signatory

Figure 7.1   Example 1 — typical descriptive consent for sewage works.
(NRA Water Quality Series No: 17 © Environment Agency, UK.)

NATIONAL RIVERS AUTHORITY

Reference

WATER RESOURCES ACT 1991 — CONSENT TO DISCHARGE

The National Rivers Authority, in pursuance of its powers under the above mentioned Act, HEREBY GIVES CONSENT to the discharge described hereunder subject to the terms and conditions set out below.

Name & Address of Applicant:	Any Water Services Co Ltd 123 High Street Anytown
Date of Application:	18 July 1992
Date of Consent:	10 November 1992
Description of Discharge:	Type: Final Sewage Effluent From: Somewhere STW To : Somewhere Stream

This consent shall not be taken as providing a staututory defence against a charge of pollution in respect of any poisonous, noxious or polluting constitutents not specified herein.

Conditions

1. General
   a. This consent shall come into force on 1 May 1993.
   b. Except with the agreement of the person making the discharge under this consent, no notice shall be served revoking the consent or modifying the conditions before 1 May 1995.
   c. For the purpose of applying the conditions identified in section 3 below, the discharger shall provide and maintain facilities which will enable the Authority's representatives to take flow measurements of the final sewage effluent which is discharged at the outlet. The discharger shall identify the facility with a clearly visible sign, distinguishing it from any other and provide a clearly visible notch, mark, or device indicating the level equivalent to the maximum instantaneous consented flow.
   d. For the purpose of applying the conditions identified in section 4 below, the discharger shall provide and maintain facilities which will enable the Authority's representatives to take discrete samples of the final sewage effluent which is discharged at the outlet. The discharger shall identify the facility with a clearly visible sign distinguishing it from any other.
   e. The discharger shall provide to the Authority's satisfaction a drawing showing the precise location of the facilities provided in accordance with conditions (c) and (d) above not later than one month prior to the date of enforcement of this consent.
   f. Facilities shall be provided for safe and convenient access to enable the Authority's representatives at any time to take samples, carry out flow measurements and inspection to ensure that the conditions are complied with.

2. As to Outlet
   An outfall shall be sited at NGR XX 9999 9999 and shall be so constructed that it is used for the discharge of final sewage effluent derived only from this sewage treatment works.

Continued overleaf

Figure 7.2 Example 2 — typical numerical sewage works consent. (NRA Water Quality Series No:17 © Environment Agency, UK.)

3   As to Discharge
   a The maximum instantaneous rate of discharge shall not exceed 1.8 litres per second.
   b The volume discharged under dry weather flow conditions shall not exceed 39.6 cubic metres in any period of twenty four hours.

4   As to Discharge Composition
   a The discharge shall:
      (i) contain no visible signs of oil or grease.
      (ii) at no time contain any matter, other than matter specifically authorized or limited by numerical conditions in this consent, to such an extent as to cause the receiving waters to be poisonous or injurious to fish or the spawning ground, spawn or food of fish.
   b In any series of samples of the final effluent taken over any twelve month period as listed in column 1 of the table set out in the annex to this schedule, then, in respect of the following determinands, no more than the relevant number as permitted in column 2 of the table shall be:
      (i) in excess of 40 milligrams per litre of biochemical oxygen demand (BOD) measured after 5 days at 20°C with nitrification suppressed by the addition of allyl thiourea;
      (ii) in excess of 60 milligrams per litre of suspended solids (measured after drying for one hour at 105°C).
      (iii) in excess of 10 milligrams per litre of ammoniacal nitrogen expressed as nitrogen.
   c No single sample of the final effluent discharged shall have:
      (i) in excess of 80 milligrams per litre of biochemical oxygen demand (BOD) measured after 5 days at 20°C with nitrification suppressed by the addition of allyl thiourea;
      (ii) in excess of 120 milligrams per litre of suspended solids (measured after drying for one hour at 105°C).
      (iii) in excess of 20 milligrams per litre of ammoniacal nitrogen expressed as nitrogen;
      (iv) a pH value less than 6 or greater than 9.

NRA Regional Office
Address   ....
                                         NRA Authorized Signatory
                                         NRA Region

Figure 7.2   Example 2 — typical numerical sewage works consent (continued).

7.5   LANDFILL

In the UK, landfill is still currently the cheapest and most widely applied technique for the disposal of wastes. More than 95% of all solid waste in England and Wales is landfilled which, if properly executed, enables low-grade land to be put into productive use at a relatively low cost. The landfilling of liquid effluents is an acceptable option in limited cases but alternative disposal routes should be considered. However, the proposed EC directive on the landfilling of waste still leaves some scope for the disposal of liquid wastes. The UK position on landfilling waste is reported in the DoE *Waste Management Paper No 26*[5].

The proposed EC landfill directive (see Section 3.11.5, page 52) as amended contains a concession to countries like the UK which practise 'co-disposal' of industrial waste with municipal waste. A period of five years is now to be allowed, after which co-disposal will be banned.

Note that controlled landfill costs are increasing rapidly due to stringent site controls being placed upon the operators as a direct result of environmental concern and legislative enactments. Also, a landfill tax has recently been introduced. These pressures will lead to a continuing reduction in the landfilling of special wastes.

Many of the on-site and off-site treatment technologies produce a residual quantity of material which cannot be utilized or discharged to air or aquatic environments, necessitating safe disposal in landfills for the foreseeable future.

The public perception of landfill disposal sites is coloured by an environmental legacy of badly designed and neglected facilities, some of which have become notorious — for example, Love Canal and Swartz Creek in the US. In the EU, a register of contaminated sites is being compiled by the European Environmental Agency and in the US, the Superfund program is pursuing a huge clean-up campaign. Advances in knowledge mean that state-of-the-art landfills now have a relatively low environmental impact. The three key elements are:

- enclosing the waste deposits underneath to prevent groundwater contamination and gas migration;
- covering the waste deposits to limit air emissions — that is, gases, odours and particulates — and rain water penetration;
- long-term stability due to subsidence, erosion and geological changes.

Figure 7.3 shows the main features of a modern landfill. In addition, there is a need to drain any liquid which collects within the waste and becomes contaminated. This drained liquid requires processing using the most appropriate treatment technologies like any other similar waste liquid stream (see Section 6.4, page 124). The other important issue is landfill gas which is generated by the degradation of organic materials. It has caused major problems in some old landfills but on modern sites the gas is collected and either flared or preferably used as a low calorific value fuel.

The ultimate shutdown of a landfill has been one of the most poorly managed areas in the life cycle of such facilities and has been the major factor in adverse environmental impact. Shutdown requires the additional sealing of the surface together with a monitoring system to alert operators to any degradation of the landfill's integrity. However, the long-term management of redundant landfills may be compromised if the operating company runs into financial difficulties.

Alternatives to landfills are deep wells which have been used in the US for the disposal of hazardous liquid wastes, and old mines — for example, salt mines in Germany and Poland. The latter offer very stable geological structures.

MANAGEMENT OF PROCESS INDUSTRY WASTE

The main future application for such deep repositories is the storage of radioactive waste. These waste storage facilities will have extremely sophisticated monitoring and control systems.

7.6 LIQUID/SOLID WASTE CONTRACTORS

Whenever a waste or potential waste has been identified, the waste producer must be aware of a duty of care for that waste until it is finally disposed of or destroyed. The UK Environmental Protection Act defines duty of care on any person who produces, carries, keeps, treats or disposes of waste. The producer has a duty to take all such reasonable steps to:
- prevent offences by any other person;
- prevent escape of the waste;

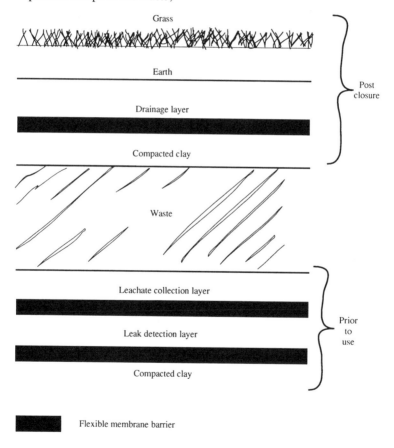

Figure 7.3   Main features of a landfill.

- transfer waste only to an authorized person;
- transfer waste with a written description for the recipient of the waste to meet the first two points above.

This duty of care commits the waste producer to being satisfied that the persons receiving the waste are authorized to do so and that the waste has been adequately identified and described (see Section 3.4.5, page 37). Auditing and monitoring of a waste disposal route is carried out prior to letting a contract to dispose, treat or destroy the particular waste. The waste producer needs to audit the waste carrier, the waste treatment facility, the subsequent disposal/recycling routes resulting from such treatment and the disposal or destruction destination of the wastes.

It is important that the waste disposal route is chosen with a particular waste stream in mind. It is not sufficient to rely on the fact that the contractor satisfactorily deals with waste A when considering the disposal of waste B which has a different composition and properties.

In the UK, major waste disposal contractors are members of the trade association, the Environmental Services Association (formerly the National Association of Waste Disposal Contractors), which produces an annual directory (see Appendix 1). The four steps in monitoring of the disposal of waste are:

(1) identification of a suitable waste disposal facility;
(2) assessment of the disposal facility management and operation;
(3) selection and assessment of a waste carrier;
(4) follow-up monitoring of a waste disposal route.

For each of these key stages, a systematic consideration of the following check-lists is made.

STEP 1 — IDENTIFICATION OF A SUITABLE WASTE DISPOSAL FACILITY
- Select a facility on its ability to handle and process waste, using methods that provide adequate safety to people and the environment and meet the statutory requirements. Discuss with the contractor and agree the appropriate batch size for the disposal and frequency of arising of the waste.
- Identify sites that are appropriately authorized to receive such waste in conjunction with the Environment Agency.
- Obtain a copy of the relevant waste management licence for each waste disposal site.
- Check the technical and financial standing of the operating company and its insurance cover.
- Visit the disposal facility prior to finalizing the contract.
- Ask to see the progress of a particular waste through the system, particularly if the company is registered to BS 5750[6] (or BS 7750[7]).

STEP 2 — ASSESSMENT OF THE DISPOSAL FACILITY MANAGEMENT AND OPERATION

On the visit to the waste disposal facility, the waste producer should be satisfied that:

- the disposal operatives are aware of the disposal principles being applied;
- the disposal company has a responsible attitude to acceptance and treatment of waste;
- full records are made of loads received and the weights of those loads;
- the site has a good safety record and there is an air of tidiness and good order;
- there are written emergency procedures;
- there are adequate firefighting facilities;
- there are adequate environmental protection facilities;
- the approach to the site is tidy and free from deposits of waste or windblown material;
- the site has adequate analytical facilities;
- the waste facility has a good relationship and acceptance in the neighbourhood;
- waste reception areas are segregated and carefully labelled.

For landfill sites check that:

- there is a controlled forward plan for site utilization;
- there is controlled placing of all hazardous loads;
- all run-off water is adequately collected, treated and discharged to a consented discharge within consent. Ask to see the results of analysis of the discharge;
- there is adequate protection against groundwater contamination. Ask to see ground water monitoring data;
- the site has a good record of operation from the waste disposal authority. Ask to see the waste inspector's reports.

For a treatment facility check that:

- the contractor is operating a sound process. Is the contractor doing what the waste producers want?
- the products of treatment are adequately dealt with — for example, solids sent to suitable landfill (subsequent landfill site should also be visited) and aqueous liquids further treated and directed to a consented discharge;
- the process and equipment are compatible with the particular waste stream destined to the facility.

For incineration facilities check that:

- pretreatment of waste prior to incineration is acceptable;
- incinerator process control is reliable and incinerator temperature records are available;

- there is no smoke or fume problem from the stack;
- stack particulates are adequately removed and disposed (further checks will be needed on the disposal of the particulates);
- there is evidence of complete combustion of wastes;
- if there is a wet scrubbing system, the aqueous streams are adequately treated prior to discharge by a consented route;
- in addition to the waste disposal licence, check that the appropriate department in the Local Authority Air Pollution Control system is satisfied with the performance of the plant.

STEP 3 — SELECTION AND ASSESSMENT OF A WASTE CARRIER
- The duty of care insists that only authorized carriers are used — check this before the waste is dispatched.
- Check that the proposed carrier is experienced in carrying the types of waste involved and in waste disposal activities.
- Does the carrier have a system of keeping records of waste loads? In particular, details relating to special waste transfers have to be kept by the carrier for a minimum period of three years.
- Check that the carrier has equipment suitable for handling the particular waste, and for the plants and sites visited.
- Establish whether the carrier's employees understand the processes and materials handled and are trained to deal with spillages and other emergencies.
- Make sure that you, as waste producer, have given sufficient information for the material to be transported safely.
- Confirm with the Environment Agency that the carrier is a bona fide operator.
- Ensure that before the waste departs from the producer's site that it is correctly labelled and is accompanied by the correct documentation.

STEP 4 — FOLLOW-UP MONITORING OF THE WASTE DISPOSAL ROUTE
To properly discharge the duty of care, the progress of the waste disposal contract should be monitored carefully by visiting the contractor at irregular intervals during the life of the contract.

Keep records of weights of materials sent to the contractor, and check against the completed certificates of disposal returned by the contractors. Investigate discrepancies promptly.

7.7 TRANSPORT OF WASTE

Castle[8] has produced a comprehensive guide to the transport of dangerous goods which includes waste substances and articles.

The duty of care requires the producer to transfer the waste with a written description. The reason for this need is to protect the health and safety of the company and contractor's employees and any other persons that may be placed at risk whilst the material is in transit or being handled at the waste disposal facility. This duty of care can be discharged by providing a hazard data sheet for the particular waste just as one would do for any goods.

The Recommendations on the Safe Transport of Dangerous Goods[9], the so-called 'Orange Book', published by the UN in Geneva is the basis for all international regulations. It sets out the proper classification, packaging, identification, labelling and documentation of dangerous goods. In the UK, the relevant statute is the Health and Safety at Work Act 1974 and the following regulations concerning transportation equally apply to waste materials:
* the Road Traffic (Carriage of Dangerous Substances in Road Tankers and Tank Containers) Regulations 1992, as amended SI 743 (RTR). The regulations require labelling of the vehicle and information to be kept in the vehicle cab;
* the Road Traffic (Carriage of Dangerous Substances in Packages etc) Regulations 1992, as amended SI 742 (PGR) has similar requirements for materials in skips or in bulk in lorries — that is, labelling of vehicles and carrying of Tremcards in vehicles;
* the Road Traffic (Training of Drivers of Vehicles Carrying Dangerous Goods) Regulations 1992, as amended SI 744 (DTR);
* the Carriage of Dangerous Goods by Road and Rail (Classification, Packaging and Labelling) Regulations (1994) SI 669(CDG-CPL) abolished the 7000 series classification and the exclamation mark diamond;
* the Carriage of Dangerous Goods by Rail Regulations (1994) (CDG Rail) introduced to cover the privatization of the railways, places responsibility on the railway companies to operate services involving dangerous goods safely. (Note — equivalent regulations are being produced covering road transport.)

There are additional regulations covering explosives and radioactive materials. Note that the EC has approved a Directive 94/55 to be adopted by all member states by 1 January 1997. It is based on a new agreement for road transport, called ADR (the initials are based on the French title). CDG-CPL has already been amended accordingly and the other regulations are being modified to ADR. These regulations require:
* the classification of dangerous substances — it is covered by UN numbers, proper shipping names, packing groups and classes. There are guidelines on

Figure 7.4 Primary and secondary hazard labels. (Castle, M., 1995, *The Transport of Dangerous Goods*, reproduced by permission of PIRA International.)

substances which are not classified by name, constitute a multiple hazard, mixtures and environmentally hazardous substances;
• the packaging of the consignment — there are four recognized types: limited quantities, up to 400 kg or 450 l, intermediate bulk containers up to 3000 kg and portable tanks. All types must be 'suitable for the purpose' and in good condition, compatible with the contents, provide sufficient expansion space for liquid contents, secure closures and inner packaging and every packaging for liquids must be leak proof tested. The testing of packaging is a major step in the introduction of any novel or modified packaging system;
• the marking and labelling of the consignment (and where necessary the vehicle) — marking places with the UN number and proper shipping name on the package, IBC (intermediate bulk container) or tank. If helpful, a further technical name can be added. Labelling covers the use of diamond shaped labels which indicate the hazard(s). These allow the hazard to be identified remotely by the design and colour. Two examples are shown in Figure 7.4. There may be more than one hazard and then a subsidiary label appears without a class number. If liquids are carried then an upright label (two arrows) is added. Figure 7.5, page 170, illustrates a drum with the appropriate marks and label. Nationally, there was a system of Hazchem labels for a whole range of materials with numbers in the '7000' series used for wastes (see Table 7.2, page 171). These have been withdrawn for packages and IBCs; withdrawal for tankers is from 31 December 1996;

MANAGEMENT OF PROCESS INDUSTRY WASTE

- emergency procedures — information must be provided with the vehicle when materials are in transit so that, should an emergency arise, action can be taken. A system of Tremcards (meaning Transport Emergency Cards, a trademark of CEFIC, the European Chemical Industry Council) has been developed. Tremcards are documents that satisfy the regulation for information in writing available in transit. An example is shown in Figure 7.6, page 172. Further advice can be obtained from the National Chemical Emergency Centre, Culham (see Appendix 1).

The information relates to the material being carried and is kept in the cab of the vehicle. The content of the information must:
- state the actions to be taken in an emergency;
- give the identity of the substance;
- give the main dangers from the materials.

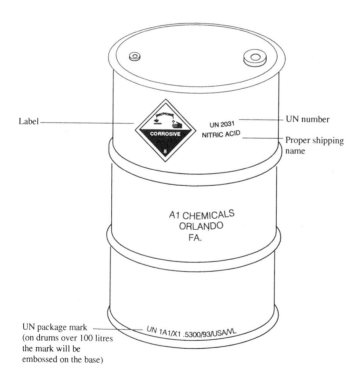

Figure 7.5 Marking and labelling for a single packaging (steel drum).

TABLE 7.2
Hazchem codes for waste substances

Name of substance	Substance identification number
Hazardous waste, liquid, containing acid	7006
Hazardous waste, solid or sludge, containing acid	7007
Hazardous waste, liquid, containing alkali	7008
Hazardous waste, solid or sludge, containing alkali	7009
Hazardous waste, flammable liquid, flash point below 21°C	7010
Hazardous waste, flammable liquid, flash point 21°C to 55°C	7011
Hazardous waste, flammable solid or sludge, not otherwise specified (n.o.s.)	7012
Hazardous waste, n.o.s., miscellaneous, packaged	7013
Hazardous waste, solid or sludge, n.o.s.	7014
Hazardous waste, liquid, n.o.s.	7015
Hazardous waste, solid or sludge, toxic, n.o.s.	7016
Hazardous waste, liquid, toxic, n.o.s.	7017
Hazardous waste, solid, containing inorganic cyanides	7018
Hazardous waste, liquid, containing inorganic cyanides	7019
Hazardous waste, solid or sludge, agrochemicals, toxic, n.o.s.	7020
Hazardous waste, liquid, agrochemicals, toxic, n.o.s.	7021
Hazardous waste, containing isocyanates, n.o.s.	7022
Hazardous waste, containing organolead compounds, n.o.s.	7023
Hazardous waste, sludge, containing asbestos	7030

TRANSPORT EMERGENCY CARD (Road)　　　　　　　　　　　　CEFIC TEC(R) - 30G30
　　　　　　　　　　　　　　　　　　　　　　　　　　　　　1995
　　　　　　　　　　　　　　　　　　　　　　　　　　　　　Class 3 ADR
　　　　　　　　　　　　　　　　　　　　　　　　　　　　　It

CARGO　FLAMMABLE LIQUIDS, HAVING A FLASH-POINT BELOW 23°C NOT TOXIC AND NOT CORROSIVE

Name of substance(s): ───────────────────────────
- Colourless liquid - Perceptible odour / Colourless paste - Perceptible odour / Coloured liquid - Perceptible odour / Coloured paste - Perceptible odour
- Completely miscible with water / Immiscible or partly miscible with water - Lighter than water / Immiscible or partly miscible with water -
- Heavier than water

NATURE OF HAZARD
- Highly flammable
- May evaporate quickly
- The vapour may be invisible. The vapour is heavier than air and spreads along ground
- May form explosive mixutre with air, particularly in empty uncleaned receptacles
- Heating will cause pressure rise, severe risk of bursting and subsequent explosion
- The vapour may have narcotic effect
- May have irritant effect: on eyes, on skin, on air passages
- Decomposes in a fire giving off toxic fumes. The effect of inhalation may be delayed+

BASIC PERSONAL PROTECTION
- Suitable respiratory protective device
- Goggles giving complete protection to eyes
- Plastic or synthetic rubber gloves. Boots
- Eyewash bottle with clean water

IMMEDIATE ACTION BY DRIVER - Notify police and fire brigade
- Stop the engine
- No naked lights. No smoking
- Mark roads and warn other road users
- Keep public away from danger area
- Keep upwind

SPILLAGE
- Stop leaks if without risk
- Use low-sparking handtools and explosion-proof electrical equipment / Use explosion proof electrical equipment
- Prevent liquid entering sewers, basements and workpits. Vapour may create explosive atmosphere
- Contain or absorb leaking liquid with sand or earth or other suitable material. Consult an expert
- Warn everybody - Explosion hazard
- If substance has entered a water course or sewer or been spilt on soil or vegetation, advise police

FIRE
- Keep container(s) cool by spraying with water if exposed to fire
- Extinguish with waterspray, foam or dry chemical
- Do not use water jet

FIRST AID
- If substance has got into the eyes, immediately wash out with plenty of water. Continue treatment until medical assistance is provided
- Remove contaminated clothing immediately and wash affected skin with soap and water
- Seek medical treatment when anyone has symtoms apparently due to inhalation or contact with skin or eyes
- Persons who have inhaled the fumes produced in a fire may not show immediate symptoms. Patient must be kept under medical supervision for at least 24 hours+
- In case of burns immediately cool affected skin as long as possible with cold water

Additional information　　　　　　　　　　　　　TELEPHONE

COPYRIGHT BY CEFIC　Prepared by CEFIC from the best knowledge available; no responsibility is accepted that the information is sufficient or correct in all cases

APPLIES ONLY DURING ROAD TRANSPORT　ENGLISH

Figure 7.6　Example of a Tremcard. (Courtesy of NCEC, AEA Technology.)

In view of the extensive changes currently under way, Rechem International Ltd has set up a free advisory service at Panteg, Pontypool, Gwent[10].

The following description applies to consigning waste materials in the United Kingdom. Other requirements are in place for transfrontier shipment of wastes (see Section 7.8). A consignment note is a record of the transfer of a quantity of material from one waste keeper to another. In general, the consignment note states how much of a specified waste is transferred from, say, the waste producer to the waste disposer, when it was transferred and where it was disposed.

When dealing with special wastes, there is a statutory set of forms on which this information is transferred and recorded. The Special Waste Regulations (1996), as well as defining special waste, require a statutory consignment note (see Appendix 5) to be used for the disposal of each consignment of special waste. They also lay down duties for 'producers' of waste, and the 'carriers' and 'disposers' — that is, operators of waste disposal facilities — of special waste.

7.8 TRANSFRONTIER SHIPMENT OF WASTE

There is a longer term aim within the European Union to apply a proximity principle which requires waste disposal to take place as close to the point of production as possible.

EC regulation 259/93 on the 'Supervision and Control of Shipment of Waste within, into or out of the EU' came into operation on 6 May 1994. The regulation replaces and extends existing EC legislation on transfrontier shipments of waste within, into or out of the EU. Its purpose is, among other things, to give effect to the Basel Convention on the control of transboundary movements of hazardous waste and its disposal. However, the regulation applies to all waste as defined in other EC legislation. The main provisions of the new regulation lay down procedures and controls for waste movements involving different groups of countries, with separate provisions made for wastes intended for disposal and those intended for recovery.

The regulation contains procedures for the notification of waste movements, and acknowledgement or objection procedures for regulatory authorities to follow. The authority in the country receiving the waste must authorize the shipment. Conditions may be set and further information requested by the authorities. The waste is also checked at the end of its journey.

Currently, hazardous waste shipments for disposal to countries outside of the Organization for Economic Co-operation and Development (OECD) are banned. Shipments can occur if they are for recycling, but this leaves a potential loophole for unscheduled dumping.

In the UK, the regulation has been implemented by the Transfrontier Shipment of Waste Regulations 1993. Similar principles to the UK consignment note apply to transfrontier consignment of wastes. Special forms are available from the Department of Environment.

In February 1995, a draft management plan on the import and export of waste was issued which will tighten regulations so that:
- all exports of waste for disposal are banned;
- OECD countries can accept waste for recycling;
- imports can be accepted for disposal of hazardous waste from developing countries which do not have facilities;
- imports can be accepted for disposal of small quantities of waste from developed countries where the required facilities would be uneconomic.

REFERENCES IN CHAPTER 7
1. *CEP Software Directory*, annual, a supplement to *Chemical Engineering Progress* (AIChE, New York, USA).
2. Hesketh, H.E. and Cross Jr, F.L., 1983, *Fugitive Emissions And Controls* (Ann Arbor Science, Butterworth-Heinemann, Oxford, UK).
3. Moretti, E.C. and Mukhopadhyday, N., 1993, *Current and Potential Future Industrial Practices for Reducing and Controlling Volatile Organic Compounds* (Centre for Waste Reduction Technologies (CWRT), AIChE, USA).
4. National Rivers Authority, 1994, Discharge consents and compliance — The NRA's approach to control of discharges to water, *Water Quality Series No 17* (NRA, Bristol, UK — now part of the Environment Agency).
5. *Waste Management Paper No 26 — Landfilling Waste*, 1986 (Department of Environment, UK).
6. BS 5750: Quality Management and Quality Assurance Standards (British Standards Institution, UK).
7. BS 7750: Specification for Environmental Management Systems, 1994 (British Standards Institution, UK).
8. Castle, M., 1995, *The Transport of Dangerous Goods: A Short Guide to International Regulations* (Pira International, Leatherhead, UK).
9. *The Recommendations on the Safe Transport of Dangerous Goods* (United Nations, Geneva, Switzerland) available from HMSO, UK.
10. Anon, 1995, Hazwaste: all change, *Hazardous Cargo Bulletin*, September, 9.

8. SAFETY AND HEALTH ISSUES

8.1 INTRODUCTION

There are some unique safety and health considerations when handling, treating and discharging waste (see Table 8.1, page 176).

Make sure that waste is categorized, segregated and labelled correctly. This eliminates, for example, the risk of water exposure to hazardous waste that is considered innocuous after being mixed inadvertently with non-hazardous waste. Pay particular attention to those points in the waste management system where the waste is transferred from one operating unit to another or to a third party, eliminating the chance of communication breakdown.

The main hazards are:
- physical — the handling and transportation of waste and associated containers and vehicles — that is, the treatment plants (confined spaces, deep tanks, low density in aerated systems, mechanical equipment, platforms, stairs and so on);
- chemical — corrosive, toxic (poison is now no longer an acceptable term) or combustible wastes, including fumes, off-gases and sludges, and treatment chemicals;
- biological — micro-organisms in the waste or developing in the treatment system.

Carry out a hazard and operability study (Hazop) of the effluent handling and treatment system to analyse systematically the causes and effects of abnormal operations. It should be updated if the effluent or process are changed. The IChemE has a number of excellent books and training packages on this topic (see Appendix 3) and Martin *et al* is a further useful source of information[1]. The UK Royal Society for the Prevention of Accidents (RoSPA) provides useful material such as posters and books on general safety and health issues.

Croner's Waste Management Guide[2] has a comprehensive section on 'waste charts' which set out for each waste material/compound the physical, chemical, medical and environmental hazards. There are also notes on handling precautions, labelling, storage and legislation.

TABLE 8.1
Overview of Chapter 8

8.2, page 176 Legal framework	• Health and Safety at Work Act • Control of Substances Hazardous to Health Regulations • Reporting of Injuries, Diseases and Dangerous Occurrences Regulations • EEC Directive 80/1107
8.3, page 177 Confined spaces	• hazard from fumes and lack of oxygen
8.4, page 177 Deep tanks	• hazard from drowning in liquids, sludges and solids
8.5, page 178 Combustible materials	• hazard from fire and explosion
8.6, page 178 Volatile organic compounds (VOCs)	• harmful organic chemicals
8.7, page 178 Anaerobic digesters	• hazard from gases generated
8.8, page 178 Biological hazards	• bacteria and viruses
8.9, page 179 Toxicity	• hazard from skin contact and ingestion • available databases
8.10, page 179 Corrosivity	• risk of damage to tissue by contact, ingestion and inhalation • available databases
8.11, page 179 Carcinogenicity	• increased risk of cancer on exposure • available databases

8.2 LEGAL FRAMEWORK

There are four main legislative items in the UK:

• Health and Safety at Work, etc Act (HSWA) 1974 sets out the responsibilities of employers and supervisors to protect the health and safety of employees, including training and advice. The more recent Management of Health and Safety at Work Regulations 1992 introduce risk assessment as a key tool;

SAFETY AND HEALTH ISSUES

- Control of Substances Hazardous to Health Regulations (COSHH) 1994 requires a formal assessment and record of the risk to a person of handling or exposure to a hazardous substance. HASTAM[3] has produced a systematic guide to performing a COSHH audit, which is also available as a computer program. The Health and Safety Commission (HSC) has published three Approved Codes of Practice (ACOPs) in a single volume covering the 1994 Regulations (General COSHH ACOP (Control of Substances Hazardous to Health)), Carcinogens ACOP (Control of Carcinogenic Substances) and Biological Agents ACOP (Control of Biological Agents). Other ACOPs being revised cover vinyl chloride, pottery production, *Legionella* and non-agricultural pesticides;
- Reporting of Injuries, Diseases and Dangerous Occurrences Regulations (RIDDOR) 1985 requires notification to the Health and Safety Executive (HSE) of certain accidents, incidents and diseases linked to specific types of work in different industries and that there are adequate protective control measures;
- EEC Directive 80/1107 (27 November 1980) requires minimization of exposure of workers to harmful physical, chemical and biological agents.

8.3 CONFINED SPACES

Deaths in confined spaces from dangerous fumes or a lack of oxygen are still far too common in the waste industry. Treat any area where natural ventilation may be inadequate to prevent the build-up of gases or vapours, or where oxygen levels can fall (due to absorption by chemicals or bacteria) as a confined space. This could include open tanks and partitions.

All entry into confined spaces should be controlled by a 'permit-to-work' system (HSE Guidance Note GS5).

The proper use of, and training in, ventilation, protective clothing and breathing apparatus is vital when working in confined spaces.

8.4 DEEP TANKS

Drowning in, or injury from falling into, deep tanks is a hazard. Consider the danger not only where liquids are being processed and stored but also for sludges and solid wastes. A particular hazard is introduced in aerated tanks when the bulk density of the liquid becomes much less than that of the human body. Minimize entry to such vessels and use interlocks to protect workers against the unscheduled inrush of material. Life belts may be appropriate but may be less effective in aerated tanks where the bulk density is low.

8.5 COMBUSTIBLE MATERIALS

The dangers of fire and explosion directly from combustible wastes and indirectly by generation of combustible gases are well known, particularly in designated storage areas. Vigilance is needed to prevent wastes from being dumped or temporarily stored in an unplanned way. BS 5345 classifies zones as:
- zone 0 — explosive gas-air mixture continuously present or for long periods;
- zone 1 — explosive gas-air mixture likely in normal operations;
- zone 2 — explosive gas-air mixture unlikely in normal operations and if it occurs for short periods only.

Eliminate ignition sources and ensure that any gas-air mixture is below 25% of the lower explosive limit.

8.6 VOLATILE ORGANIC COMPOUNDS

Harmful organic chemicals may be released into areas where exposure to man can occur and comes under the COSHH Regulations. In addition to considering the immediate area, also be alert to possibilities of VOCs travelling and accumulating at some point remote from the equipment — for example, leakage through underground drainage system or into ventilation ducts.

8.7 ANAEROBIC DIGESTERS

Anaerobic digesters present a unique combination of hazards. These devices generate gases that are combustible, toxic, anoxic and corrosive. Leakages of gas out of such systems and of air into them are prevented by proper design, operation and maintenance. Digesters which are heated pose an additional hazard.

8.8 BIOLOGICAL HAZARDS

Refer to the recent COSHH Code of Practice on biological agents (see Section 8.2, page 176). Infections via cuts and abrasions can occur from bacteria. Tetanus can be caught from a wide range of materials such as soils and dusts while leptospirosis is of concern in aqueous environments contaminated with infected rat or cattle urine. Immunization against tetanus is widely recommended but further immunization is normally reserved for particular risk areas.

In wastewaters and treated waters, Legionnaires disease must be considered, particularly if warm waters are reused[4].

Sludges, particularly sewage, can pose a major hazard due to bacteria and viruses such as poliomyelitis, cholera and typhoid.

8.9 TOXICITY

A waste's toxic effect by skin contact and ingestion should be assessed. There are major reference databases on toxicity such as *The Register of Toxic Effects of Chemical Substances*[5] produced by the US National Institute of Occupational Safety and Health. The data are presented as a lethal dose value — either LD_{50} (50% of the test animals die) or LD_{LO} (lowest published) — and the lowest value available is used in calculating the limiting concentration to define special waste. The HSE produces an Occupational Exposure Limits EH40 annual on toxicity levels for working conditions which is available in both hard copy and database format.

A new National Centre of Environmental Toxicology has been established at WRc (see Appendix 1) to provide advice and information. The Environmental Agency also operates the Toxic and Persistent Substances Centre (TAPS).

8.10 CORROSIVITY

The risk of damage to tissue by ingestion, inhalation and contact is also of concern. Clearly, the level of protective clothing and breathing equipment is dictated by the risk: the criterion is based on 15 minutes contact to cause serious damage. The HSE Labelling Regulations list substances which are corrosive.

8.11 CARCINOGENICITY

The increased risk of cancer by exposure to various materials is a complex issue. For some compounds, such as benzene, the scientific evidence for a link is well established. However, for many materials, particularly where there is a mixture of compounds, the data are either not available or inconclusive. A recent COSHH Code of Practice on carcinogenic substances should be consulted (see Section 8.2, page 176).

REFERENCES IN CHAPTER 8
1. Martin, W.F., Lippitt, J.M. and Prothero, T.G., 1992, *Hazardous Waste Handbook for Health and Safety*, 2nd edition, (Butterworth-Heinemann, UK).
2. *Croner's Waste Management Guide*, 1991 (Croner Publications, UK).
3. HASTAM, 1991, *COSHH Audit* (Mercury Books, London, UK).
4. *Legionella — A BEWA Guide*, 1993 (British Water, UK).
5. *The Register of Toxic Effects of Chemical Substances (RTECS)*, (US National Institute of Occupational Safety and Health).

APPENDIX 1 — USEFUL CONTACTS

• The Environmental Helpline	Free up-to-date information on environmental issues, legislation and technology as part of the Government's Environmental Technology Best Practice Programme (ETBPP) initiative. A specialist will work on the query for up to two hours free of charge. For small businesses, the direct help of an environmental counsellor may be offered. The ETBPP concentrates on waste minimization and cleaner technology promoting good, new and future practices. One industrial Club already launched is the West Midlands Minimization Project. Tel: + 44 800 585794, Fax: + 44 1235 463804
• Department of the Environment	2 Marsham Street, London, SW1P 3EB Tel: + 44 171 276 3000
• Environment Agency (HQ)	Rio House, Waterside Drive, Aztec West Almondbury, Bristol, BS12 4UD Tel: + 44 1454 624400, Fax: + 44 1454 624409
• Scottish Environment Protection Agency (HQ)	Erskine Court, The Castle Business Park Stirling, FK9 4TR, Scotland Tel: + 44 1786 457700, Fax: + 44 1786 446885
• Department of Trade and Industry	Information on waste management and minimization Ashdown House, 123 Victoria Street, London, SW1E 6RB Tel: + 44 171 215 5000
• Health and Safety Commission and Executive	HSE infoline Tel: + 44 541 545500
• Royal Society for the Prevention of Accidents	Cannon House, Priory Queensway, Birmingham, B4 6BS Tel: + 44 121 200 2461, Fax: + 44 121 200 1254

• Environmental Services Association (formerly National Association of Waste Disposal Contractors)	Mountbarrow House, 6–20 Elizabeth Street, London, SW1W 9RB Tel: + 44 171 824 8882, Fax: + 44 171 824 8753
• Institute of Environmental Management	58/59 Timber Bush, Edinburgh, EH6 6QH Tel: + 44 131 555 5334, Fax: + 44 131 555 5217
• Institute of Environmental Assessment	Limekiln Way, Lincoln, Lincolnshire, PE23 4DB Tel: + 44 1790 763613, Fax: + 44 1790 763630
• Waste Management Industry Training and Advisory Board	Issues certificates of technical competence, PO Box 176, Northampton, NN1 1SB Tel: + 44 1604 231950, Fax: + 44 1604 232457
• Waste Management Information Bureau	A comprehensive library, database and information service, operated by AEA Technology plc at the National Environmental Technology Centre, F6 Culham, Abingdon, OX14 3DB Tel: + 44 1235 463162, Fax: + 44 1235 463004 Email : wmit@aeat.co.uk
• Energy Technology Support Unit (ETSU)	Extensive information on utilization of waste as an energy source, operated by AEA Technology plc, ETSU, Harwell, Didcot, Oxfordshire, OX11 0RA Tel: + 44 1235 432450, Fax: + 44 1235 432923
• WRc	Aqualine database, Instrument Evaluation Service Club and National Centre for Environmental Toxicology at Medmenham, PO Box 16, Marlow, Buckinghamshire, SL7 2HD Tel: + 44 1491 571531, Fax: + 44 1491 579094
• UK Environmental Law Association	Information on law firms specializing in environmental and waste issues Honeycroft House, Pangbourne Road, Upper Basildon, RG8 8LP Tel: + 44 1491 671631
• National Chemical Emergency Centre	Offers 24 hour emergency advice service and supports CEFIC's Tremcards operated by AEA Technology plc, Culham, Abingdon, OX14 3DB Tel: + 44 1235 463060, Fax: + 44 1235 463070

APPENDIX 2 — SOFTWARE

This is a rapidly changing field and the AIChE produces an annual software directory as a supplement to *Chemical Engineering Progress* which includes waste management. The following list is indicative of the range of software available and is not intended to be comprehensive or offer any form of endorsement.

EXPERT SYSTEMS
- X-Spurt (EA Technology, Chester, UK) — advice on effluent treatment.
- Epselon (AEA Technology plc, Harwell, UK) — selection and design of effluent treatment.

DATABASES
- Clean Process Advisory System (CWRT, AIChE) — information and assistance with conceptual design stage.
- Separation Processes Service (AEA Technology plc, Harwell, UK) — separations technologies on-line manuals, selection and design guides.
- Effluent Processing Club (AEA Technology plc, Harwell, UK) — liquid effluent treatment technologies, on-line manuals, applications and selection guides.

PROCESS SIMULATORS
- PRO II (Simulation Sciences Inc, California, USA)
- Aspen Plus (AspenTech Inc, Cambridge, USA)
- Hysim (Hyprotech Inc, Calgary, Canada)
- Chemcad III (Chemstations Inc, Houston, USA)
- EnviroPro Designer (Intelligen Inc, NJ, USA)
- Environmental Simulation Program (Oli Systems Inc, USA and Davy E&E, Cleveland, UK)
- METSIM (Proware, Tucson, USA)

DISPERSION MODELLING
- BREEZE (Trinity Consultants Inc)

INTERNET SOURCES
Kumar, A. and Manocha, A., 1995, A review of environmental sites on the World Wide Web, *Environmental Progress*, 14 (3): A10–11.

APPENDIX 3 — BIBLIOGRAPHY

DTI, *Environmental Contacts, A Guide For Business* (DTI Environment Publications, ADMAIL 528, London, SW1W 8YT, UK).

HMIP, *HMIP Bibliography*, regularly updated details of all Inspectorate and some DoE publications.

DoE, *Environmental Facts, A Guide to using Public Registers of Environmental Information*, includes Register of Current Waste Management Licences, Register of Waste Carriers, Register of Premises on which Waste that Contains Explosives is Kept and IPC Register, EP335.

DoE, *DoE Digest of Environmental Protection and Water Statistics*, annual report.

1995, *Environmental Technology Market Sourcebook* (Frost & Sullivan Market Intelligence Report).

Scott, J.S. and Smith, P.G., 1980, *Dictionary of Waste and Water Treatment* (Butterworth, London, UK).

Lees, N. and Woolston, H. (eds), 1992, *Environmental Information, A Guide to Sources*, (The British Library).

Sharratt, P. (ed), 1995, *Environmental Management Systems* (IChemE, Rugby, UK).

Hills, J.S., 1995, *Cutting Water and Effluent Costs* (IChemE, Rugby, UK).

DIRECTORIES
- *Environment Business Directory* — Tel: 0181 877 9130
- *Waste Management Yearbook* — Tel: 01279 442601
- *Aspinwalls Sitefile Directory* — Tel: 0181 810 9979
- *Directory of Environmental Consultants* — Tel: 0171 278 7624
- *Waste Recycling & Environmental Directory* — Tel: 0171 987 6999
- *Environment Industry Yearbook* — Tel: 01225 330312
- *ESA Membership Yearbook and Directory* — Tel: 0171 824 8882

MANAGEMENT OF PROCESS INDUSTRY WASTE

- *CEP Software Directory* — annual supplement to *Chemical Engineering Progress*, published by AIChE and available from IChemE, Rugby, UK.
- *List of Consultants* — IChemE, Rugby, UK.

JOURNALS
- *Waste Management* — Journal of Institute of Waste Management
- *The Waste Manager* — Journal of Environmental Services Association
- *Environmental Law Monthly* (Monitor Press)
- *Environment Newsletter* (CBI)
- *Water and Environment Management* (Thomas Telford)
- *Water and Waste Treatment* (Faversham House Group)
- *Process Safety and Environmental Protection* (IChemE)
- *Environmental Protection Bulletin* (IChemE)

APPENDIX 4 — GLOSSARY OF WASTE DISPOSAL TERMS

A brief glossary of terms is included but a comprehensive listing can be found in the *Lexicon of Environmental Terms* (UK Environmental Law Association).

Anaerobic	process requiring the absence of, or not dependent on, the presence of free oxygen or air
Aerobic	process requiring free oxygen or air
Anoxic	lack or absence of oxygen
AWMA	Air & Waste Management Association, USA
BAT	best available techniques
BATNEEC	best available techniques not entailing excessive cost
BEP	best environmental practice
BOD(5)	biological/biochemical oxygen demand (5 day)
BPEO	best practical environmental option
BPM	best practical means
Broker	arranger of transfer of waste
COD	chemical oxygen demand
Co-disposal	the disposal of mixed wastes together (commonly domestic, commercial and industrial, usually in landfill and often including liquids and sludges)
Competent authorities	the regulatory authorities of concerned countries (ref: OECD, Basle)
Consignee	the person to whom the wastes are shipped
Controlled waste	household, industrial and commercial waste or any such waste (EPA S75).
COPA	Control of Pollution Act (1974) (see also regulations issued under this act)
COPA (Amendment)	amended 1989
COSHH	Control of Substances Hazardous to Health Regulations 1988
CWRT	Centre for Waste Reduction Technologies, AIChE, USA

DoE	Department of the Environment
DTA	direct toxicity assessment
EA	Environment Agency
EC	European Commission
EHO	environmental health officer
EMAS	eco-management and audit scheme
EMS	environmental management system
EPA	Environmental Protection Act ,1990, also Environmental Protection Agency (USA)
EPSRC	Engineering and Physical Sciences Research Council, UK
ESA	Environment Services Association
ETBPP	Environmental Technology Best Practice Programme
EU	European Union
Fly-tipping	the illegal deposit or disposal of waste
Hazop	hazard and operability study
Holder	person having possession or legal control of waste for OECD or Basel Convention
HMIP	Her Majesty's Inspectorate of Pollution (now absorbed into the Environment Agency)
HMSO	Her Majesty's Stationery Office, London
HSEC	Health & Safety Executive and Commission
HSWA	Health & Safety at Work Act 1974
Incineration	the burning of waste materials under controlled conditions in a plant specifically designed for the operation
IPC	Integrated Pollution Control, Part I Environmental Protection Act
ISWA	International Solids Waste Association, USA
IWM	Institute of Waste Management, UK
IWNC	International Waste Notification Code

LAAPC	Local Authority Air Pollution Control processes
Landfill	a method of disposal of wastes involving the deposit of the materials in or on the ground
Landfill gas	a mixture of methane and carbon dioxide which may be generated by the wastes in a landfill site
LAWDC	Local Authority Waste Disposal Company
Leachate	the contaminated liquors from a landfill site
Mono-disposal	the disposal of a single waste or type of waste on its own
NIMBY	not in my back yard
Notifier	see 'Holder'
NO	oxides of nitrogen — any mixture of NO_2, NO and N_2O
NRA	National Rivers Authority, UK (now absorbed into the Environment Agency)
OECD	Organization for Economic Co-operation and Development
OFWAT	Office of Water Services, UK
OSHA	Occupational Safety & Health (Administration or Act), USA
PCBs	polychlorinated biphenyls
pH	measure of acidity (low pH)/alkalinity (high pH) of water (pH = 7 is neutral)
Recovery faculty	an entity which is authorized to operate under applicable domestic law in the importing country to receive wastes and to perform recovery operations on them
Registered carrier	a person or company who is registered under the Control of Pollution (Amendment) Act 1989 to transport controlled waste
RIDDOR	the Reporting of Injuries, Diseases and Dangerous Occurrences Regulations 1985
RPB	River Purification Board, Scotland (now absorbed into the Scottish Environment Protection Agency)

RHA	Road Haulage Association, UK
ROSPA	Royal Society for the Prevention of Accidents, UK
SARA	Superfund Amendments and Reauthorization Act, USA
SEPA	Scottish Environmental Protection Agency
SHE	safety, health and environment
Site licence	see 'Waste management licence'. The licence refers specifically to a location and imposes various operational conditions, including details of material acceptable for disposal at the site (COPA Sec 5).
SOx	any mixture of SO_2 and SO_3
Special waste	controlled waste of any kind that may be considered so dangerous or difficult to treat, keep or dispose that special provisions are required (COPA Sec 5 EPA S62)
TE consent	trade effluent consent
Treatment	the carrying out of any operation on waste materials to alter them from their initial form and render them safer and/or easier to dispose of
TSS	total suspended solids
VOC	volatile organic compound
WAMITAB	Waste Management Industry Training and Advisory Board, UK
Waste management licence	a licence to carry out waste disposal activities issued by the WDA, WDA licence or WRA
WDA	Waste Disposal Authority, UK
WEF	Water Environment Federation, USA
WMA	Waste Management Association, USA
WRA	Waste Regulation Authority, UK (now absorbed into the Environment Agency)
WRAP	waste reduction always pays (Dow Chemical Company initiative)

APPENDIX 5 — SPECIAL WASTE CONSIGNMENT NOTE SYSTEM

The purpose of the consignment note procedure is to locate and allow regulators to trace special wastes from 'the cradle to the grave'. Thus, from producer through transport to final disposal, special wastes are located through the use of multi-copy consignment notes and are identified by a six digit code (see Table 3.4, page 29). The Special Waste Regulations 1996 introduces three alternative procedures:
- standard procedure;
- simplified procedure;
- procedure for carrier's rounds.

The material used for the sheets of the consignment note form is a carbonless copy paper, thus information written on to any of these sheets is copied on the sheets below (see Figure A5.1, page 192). An overview of the standard procedure is given in Figure A5.2, page 193.

The simplified procedure applies when several identical waste consignments are sent from one waste producer to one disposal site or recycling plant over a one year period. After using the standard procedure for the first consignment, there is no need to prenotify the EA and only four copies are needed for all subsequent movements. This procedure can also be used where off-spec products are sent back to their original supplier, for movements between two sites of the same company and where consignments are moved between a ship and a site outside the port area.

The procedure for the carrier's rounds aims to cut the amount of forms needed by the waste producers where one carrier routinely collects similiar consignments over a year from a number of producers and all the movements go to the same disposal site or recycling plant. Again, after the first round, only four copies are needed but in addition a schedule (see Figure A5.3, page 194) is added to each copy. The carriers must also prepare extra copies of the note and the schedule must be completed — part C by the carrier and part D by the producer. On delivery to the disposal or recycling site, as in the standard procedure, three copies of the note and schedule are given to the site operator. Finally, within 24 hours of collecting the first consignment, the carrier must deliver the waste to the site operator.

MANAGEMENT OF PROCESS INDUSTRY WASTE

Special Waste Consignment Note Example

FORM OF CONSIGNMENT NOTE

SPECIAL WASTE REGULATIONS 1996   Consignment Note No. _____
No. of prenotice (if different) _____   Sheet   of

A CONSIGNMENT DETAILS
PLEASE TICK IF YOU ARE A TRANSFER STATION ☐

1. The waste described below is to be removed from (name, address and postcode)
2. The waste will be taken to (address & postcode)
3. The consignment(s) will be   one single ☐   a succession ☐   carrier's round ☐   other ☐
4. Expected removal date of first consignment: last consignment:
5. Name   On behalf (of company)
   Signature   Date

6. ☏   7. The waste producer was (if different from 1)

B DESCRIPTION OF THE WASTE   No. of additional sheet(s) ☐

1. The waste is   2. Classification
3. Physical Form: Liquid ☐   Powder ☐   Sludge ☐   Solid ☐   Mixed ☐   4. Colour
5. Total quantity for removal   quantity   units (eg kg/ltrs/tonnes)   Container type, number and size
6. The chemical/biological components that make the waste special are:

Component	Concentration (% or mg/kg)	Component	Concentration (% or mg/kg)

7. The hazards are:
8. The process giving rise to waste is:

C CARRIERS CERTIFICATE
I certify that I today collected the consignment and that the details in A1, A2 and B1 above are correct. The Quantity collected in the load is:

Name   On behalf of the (company) (name & address)
Signature   Date   at   hrs.
1. Carrier registration no./reason for exemption   2. Vehicle registration no. (or mode of transport, if not road)

D CONSIGNOR'S CERTIFICATE
I certify that the information in B and C above are correct, that the carrier is registered or exempt and was advised of the appropriate precautionary measures.

Name   On behalf of (company)
Signature   Date

E CONSIGNEE'S CERTIFICATE

1. I received this waste on   at   hrs.   2. Quantity received   quantity   units (eg/kg/ltrs/tonnes)
3. Vehicle registration no.   4. Management Operation
I certify that waste management licence/ authorisation /exemption no.   authorises the management of the waste described in B

Name   on behalf of (company)
Signature   Date

Figure A5.1   Special waste consignment note example. (© HMSO, UK.)

The consignment note procedure places responsibilities and duties upon the producer and successive handlers of the special waste and also the disposal site's EA office. It is the duty of the producer and each subsequent handler of the special waste to complete and return the appropriate sheet to the correct recipient. A register of consignment notes needs to be kept by the producer and each subsequent handler of the special waste and it is the duty of the disposal site's EA office to collate and record information from the consignment note system. If the EA office is not satisfied about any aspect of the notification of the waste movement or disposal, then it will use the consignment note procedure to trace its history and treatment/disposal.

THE SPECIAL WASTES REGULATIONS 1996

STAGE 1: INTENTION TO DISPOSE OF WASTE

The Waste Producer is responsible for completing PARTS A & B of each of the five copies of the prescribed consignment note/form. The five sheets are carbonised.

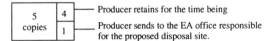

5 copies
- 4 — Producer retains for the time being
- 1 — Producer sends to the EA office responsible for the proposed disposal site.

Waste may not be moved more than 1 MONTH or less than 3 DAYS after receipt by the Environment Agency.

STAGE 2: AT THE TIME OF REMOVAL OF THE WASTE

(a) The Remover(Carrier) completes PART C on the remaining 4 copies.

(b) The Producer(Consignor) completes PART D on the remaining 4 copies.

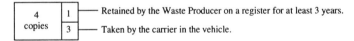

4 copies
- 1 — Retained by the Waste Producer on a register for at least 3 years.
- 3 — Taken by the carrier in the vehicle.

STAGE 3: ARRIVAL AT THE DISPOSAL SITE

The Disposal Site Operator (Consignee) completes PART E of the remaining 3 copies.

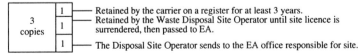

3 copies
- 1 — Retained by the carrier on a register for at least 3 years.
- 1 — Retained by the Waste Disposal Site Operator until site licence is surrendered, then passed to EA.
- 1 — The Disposal Site Operator sends to the EA office responsible for site.

Figure A5.2 Standard consignment note system.

EXAMPLE OF SPECIAL WASTE CARRIER'S SCHEDULE

SPECIAL WASTE REGULATIONS 1996: CARRIER SCHEDULE

Consignment Note No.
Sheet of

Name and address of premises from which waste was removed	I certify that today I collected the quantity of waste shown from the address given here and will take it to the address given in A2 on the consignment note	
	Quantity of waste removed	Carrier's signature and Date
	I certify that the waste collected is as detailed above and conforms with the description given in B on the relevant consignment note	
Consignment Note No	Name of Consignor	Signature and Date

Name and address of premises from which waste was removed	I certify that today I collected the quantity of waste shown from the address given here and will take it to the address given in A2 on the consignment note	
	Quantity of waste removed	Carrier's signature and Date
	I certify that the waste collected is as detailed above and conforms with the description given in B on the relevant consignment note	
Consignment Note No	Name of Consignor	Signature and Date

Name and address of premises from which waste was removed	I certify that today I collected the quantity of waste shown from the address given here and will take it to the address given in A2 on the consignment note	
	Quantity of waste removed	Carrier's signature and Date
	I certify that the waste collected is as detailed above and conforms with the description given in B on the relevant consignment note	
Consignment Note No	Name of Consignor	Signature and Date

Figure A5.3 Example of special waste carrier's schedule. (© HMSO, UK.)

THE SPECIAL WASTE CONSIGNMENT NOTE PROCEDURE

PRENOTIFICATION TO THE DISPOSAL SITE'S EA OFFICE

The purpose of the top copy of the set of five, is to prenotify the disposal site's EA office of a special waste which is to be disposed of within their area. The waste producer must give notice of not less than three working days and not more than one month of the intention to move a consignment of special waste. It is good practice to give the EA office at least one week's notice and also to send a photocopy to the receiving waste disposal site.

The waste producer completes Parts A and B of the form. The information required is clearly stated on the form and it is important to remember that variations in the type of waste can occur, together with possible contaminants. The maximum penalties for incorrect statements include a £2000 fine and/or imprisonment.

If the waste is not moved, for whatever reason, within one month of the notice, then the authority to move it is cancelled and another consignment note must be raised. Note that bank holidays, public holidays, Saturdays and Sundays do not count if they fall at the beginning or end of the month's notice. In effect, the prenotification period must begin and end on a working day.

TRANSPORTING THE WASTE

Part C of the note is completed by the driver for the transporter/contractor of the waste after checking that the details of the actual waste load are the same as in Parts A and B. It is very important that any discrepancies between the document and the load are shown on the note — if the consignment note document states '20 × 40 gallon drums' and there are only 18 such drums, then Part C must state 'only 18 drums collected'.

It is acceptable to revise the amount of material actually collected down from the amount prenotified. However, the amount prenotified is the maximum amount that can be taken off site and this prenotified amount must not be exceeded.

The responsibility of the manager of the plant which produces the waste is to ensure that:
- the carrier is made aware of the precautionary measures to be taken in handling the waste;
- there are arrangements to check that the hazard warning signs, Tremcards and Hazchem labels on the vehicle or load are appropriate;
- a responsible person, conversant with the waste, completes part D of the form and signs it;

- there is a procedure to remove, separate and register one copy, the producer's copy and that three copies are taken by the carrier with the waste.

This fulfils the requirements that a register must be kept at each premises producing special waste. Registers are kept for a minimum of three years and are made available for inspection by the EA upon request.

DISPOSAL OF THE WASTE

When the carrier arrives at the disposal point, the driver passes over all three remaining parts of the form to the waste disposer. The waste disposer checks the details of the load against the documents and then completes Part E of the form. The responsibilities of the waste disposer are to:
- enter details of the load into the record of deposits as required by the site licence;
- detach one of the three copies, the carrier's copy, and hand it back to the carrier's driver, who will take it back to the employer to be kept in the carrier's register for at least three years;
- detach another copy, the disposal site's EA office copy for their records. This must be sent no later than one working day. As for prenotification, public holidays and weekends are excluded.
- keep the final copy for the disposer's consignment note register as a record of the wastes received and deposited, cross referenced to the record of deposits. When the operation of the site ceases, this register is returned to the EA office for retention in perpetuity.

MONITORING PROGRESS

To properly discharge the duty of care, monitor the progress of the waste disposal contract by visiting the contractor at irregular intervals during the life of the contract. Keep records of weights of material sent to the contractor and make checks with the completed certificates of disposal returned by the contractors. Investigate discrepancies promptly.

APPENDIX 6 — WASTE MANAGEMENT PAPERS PUBLISHED BY HMSO

The following papers have been published by Her Majesty's Stationery Office since 1976 and are available at the HMSO Publications Centre, PO Box 276, London, SW8 5DT, UK. The titles have been rearranged into an alphabetical topic listing with the number of the paper at the end.

Arsenic Bearing Wastes — recovery, treatment and disposal; 1978 (No 20)

Asbestos Waste — arisings and disposal; 1979 (No 18)

Cadmium Bearing Wastes — arisings, treatment and disposal; 1984 (No 24)

Clinical Wastes — arisings, treatment and disposal; 1983 (No 25)

Halogenated Hydrocarbon Solvent Wastes from Cleaning Processes — reclamation and disposal; 1976 (No 9)

Halogenated Organic Wastes — arisings, treatment and disposal; 1978 (No 15)

Heat Treatment Cyanide Wastes — arisings, treatment and disposal; second edition 1985 (No 8)

Landfill Completion — a technical memorandum providing guidance on assessing the completion of licensed landfill sites; April 1994 (No 26A)

Landfill Gas — guidance on the monitoring and control of landfill gas; second edition 1991 (No 27)

Landfilling Wastes — the disposal of wastes on landfill sites; 1986 (No 26)

Licensing of Waste Management Facilities Guidance on the Drafting of Waste Management Licences — April 1994 (No 4)

Local Authority Waste Disposal Statistics 1974/75 — 1976 (No 10)

Local Authority Waste Disposal Statistics 1974/75 to 1977/78 — 1978 (No 22)

Mercury Bearing Wastes — storage, handling, treatment, disposal and recovery; 1977 (No 12)

Metal Finishing Wastes — arisings, treatment and disposal; 1976 (No 11)

Mineral Oil Wastes — arisings, treatment and disposal; 1976 (No 7)

Pesticide Wastes — arisings and disposal; 1980 (No 21)

Pharmaceuticals, Toiletries and Cosmetics — arisings, treatment and disposal; wastes from the manufacture of 1978 (No 19)

Polychlorinated Biphenyl (PCB) Wastes — reclamation, treatment and disposal; 1976 (No 6)

Recycling — guidance to local authorities on recycling; 1991 (No 28)

Relationship between Waste Disposal Authorities and Private Industry; 1976 (No 5)

Solvent Wastes (excluding Halogenated Hydrocarbons) — reclamation and disposal; 1977 (No 14)

Special Wastes — guidance on their definition; 1981 (No 23)

Tanning, Leather Dressing and Fellmongering — recovery, treatment and disposal; wastes from 1978 (No 17)

Tarry and Distillation Wastes and other Chemical Based Wastes — arisings, treatment and disposal; 1977 (No 13)

Waste Disposal Plan — guidelines for the preparation of a plan; 1976 (No 3)

Waste Disposal Surveys; 1976 (No 2)

Waste Treatment And Disposal — a review of options; guidance on the options available for; second edition 1992 (No 1)

Wood Preserving Wastes — arisings, treatment and disposal; 1980 (No 16)

INDEX

A

abatement notices	44
abatement orders	44
acid dewpoints	79, 149
activated sludge equipment	126
adsorption	130
agreements to discharge	45
air, discharges to	157–159
air pollution	34
air strippers	130
anaerobic digesters	126, 178
analyses, life cycle	96
analysis, continuous on-line	82
analytical procedures	63
analytical tests	78
assessment,	
disposal facility management and operation	166
environmental impact	31
risk	10
stream	70
waste	65–78
audit, waste	165
audits	98
environmental	8
authorized persons	37

B

bacteria	178
basic approach	90–92
batch processes	92
BATNEEC (see best available techniques not entailing excessive cost)	
best available techniques not entailing excessive cost (BATNEEC)	12, 32, 35–36
best practicable environmental option (BPEO)	12, 32, 102
biodegradability	81
bio-filters	126
biological/biochemical oxygen demand (BOD)	80
biological hazards	178
BPEO (see best practicable environmental option)	
bulk density	79

C

calorific value	81
carcinogenicity	81, 179
carriage of waste	44–45
carrier's rounds	191
CEILIC (see Environmental Impairment Liability Insurance Facility)	
cement kilns	155
centrifuges	132
Certificate of Technical Competence (COTC)	42
changes, product	96
changes in input materials	95
characterization and analysis	78
characterization parameters	78
check-lists	15–17, 39
Chemical Oxygen Demand (COD)	80
chemical reactions	68
chlorinated species	155
civil action	47
civil liability	53

clarity of liquid	80	digestion	124
cleaning	68	dioxins	155
club approach	98	directives	50
colour	80	discharge	
combustible materials	178	agreements to	45
combustion	134	consent to	45
combustion gases, cooling of	155	controlled waters	46
combustion temperatures	153	industrial liquid waste	
commercial waste	26	to sewer	45
common law	21, 47–48	to air	157–159
composting	126	to the aqueous environment	159
concentration	124	to land	159–162
conditioning	124	to sewer	45–46, 159
conductivity	80	to watercourses	159
confined spaces	177	dispersion	157
consent to discharge	45	disposal methods	17
conservation, water	92	disposal to sea	159
consignment note	173	disposal of waste	196
continuous on-line analysis	82	drainage systems	77
contracts	48	drained liquid	163
controlled waste	3, 26, 37	droplets	110
conviction	42	drowning in deep tanks	177
cooling of combustion gases	155	dryers	132
corporate environmental reports	9	duty	35
corrections	98	duty of care	12, 37–38, 39, 168
corrosivity	81, 179		
cost-benefit analysis	10		
costing structure	10	**E**	
COTC (see Certificate of Technical Competence)		EC legislation	48–57
		Eco-Audit and Management Regulation	89
cradle to grave	12, 37	Eco-Management and Audit Scheme (EMAS)	9
D		electro-dialysis	130
dangerous fumes	177	electrolysers	128
dangerous goods, transport of	84	element/compound composition	79
decisions	50	EMAS (see Eco-Management and Audit Scheme)	9
deep tanks	177		
deep-well injection	132	emissions	
deep wells	163	fugitive	158
definitions of waste	24–31	isokinetic gaseous	83
density	79	EMS (see Environmental Management System)	
dewpoint	79		

INDEX

'end-of-pipe' technology	101
environmental impact assessment	31
Environmental Impairment Liability Insurance Facility (CEILIC)	10
environmental	
investments	11
management systems	7, 89
Environmental Management System (EMS)	9
Environmental Protection Act 1990 (EPA 1990)	12
EPA 1990 (see Environmental Protection Act)	
equalization	130
evaporation	132
evaporative losses	65
exemptions	42
'expert witness'	23
explosions	178

F

fault-based liability	56
filtration	132
fires	77, 178
waste generated	159
fire water	159
fit and proper person	42
flash point	81
flocculators	130
flotation	132
fraud	47
fugitive emissions	158
fugitive losses	65

G

gas, landfill	163
gaseous treatment technologies	105–124
gas purification	110
guidance notes	36–37

H

hazard data sheets	168
hazard and operability study (Hazop)	175
hazardous waste	51
incineration	52
health	43
human	42
hydro-cyclones	132

I

incineration	134, 135–156
at sea	159
developments	155
of hazardous waste	52
incinerators,	
different types of	153
operation	53
industrial waste	26
injunction	44
in-situ systems	82
insurance	9
Integrated Pollution Control (IPC)	12, 32
interim storage on-site	16
investments, environmental	11
ion-exchange	126
IPC (see Integrated Pollution Control)	
isokinetic gaseous emissions	83

L

labelling	169, 84
lack of oxygen	177
lagoons	126
land	
discharges to	159–162
non-natural use of	47
landfill	18, 162–164
shutdown	163
gas	163
landfilling	52

land use planning	31
land waste	12
law	12
common	21, 47–48
US environmental	57–61
waste	12
leaks	159
legal framework	176
legal practices	23
Legionnaires disease	178
legislation	12
EC	48–57
lethal dose value	179
liability	
civil	53
fault-based	56
strict	56
licensing, waste management	38–43
life cycle analyses	96
liquid, drained	163
liquid treatment technologies	124–134
liquid waste contractors	164–168
local authority air pollution control	34
losses	
evaporative	65
fugitive	65

M

Magistrate's court	44
management	
solid waste	7
waste	4
management system, environmental	89
managers	
plant	11
safety	77
waste	11
marking	84, 169
mass balance	70, 92
merchant operators	150
mines	163
minimization, waste	14–16

mix wastes	17
molecular size	124
molten metal technology	134
monitoring	78–84
of waste disposal route	167

N

negligence	48
neutralizers	128
non-natural use of land	47
nuisance	43, 47

O

offence	35
off-line analysis	78
off-site waste disposal	18–19
oil/water separators	130
on-site waste treatment	16–18
Orange Book	168
organization	89–90
oxidation equipment	128
oxygen, lack of	177

P

packaging	53, 169
paint	69
particle size	124
particulate collection equipment	112
particulates	108
PFD (see process flow diagram)	
pH	80
planning permission	31
plans, waste disposal	51
plant manager	11
'polluter pays'	10, 51
pollution	
air	34
control of	12
control and waste	
regulation	31–44

precipitators	128
prescribed process	13, 46, 63
prescribed specific substances	32, 65
priorities and targets	97–98
process characterization	63–65
processes	
batch	92
semi-batch	92
process flow diagram (PFD)	65
process surveys	77
process and waste characterization	13–14
product changes	96

R

radioactivity	81
raw materials	65
records	84
recovery of residues	15–16
recycle of residues	15
recycling	97
reduction equipment	128
registration of waste carriers	44
regulations	21
regulatory authorities	12–13, 48
representative sampling	77
residues	
recovery	15–16
recycle	15
Responsible Care	7, 89
reverse osmosis	130
reviews, audits and corrections	98
risk, assessment	10
rotary kilns	154
rotating bio-contactors	126
routine monitoring	63

S

safety managers	77
sampling	78
scrap equipment and plant	135
screening systems	132

sea	
disposal to	159
incineration at	159
sedimentation	132
selection	
of the disposal method	17
of particulate collection equipment	12
of treatment methods	102–105
semi-batch processes	92
sewage sludge	159
sewer	
discharges of industrial liquid waste	45
discharges to	45–46, 159
shipment	52
shutdown of a landfill	163
sludge	124
sewage	159
solids compressed density	79
solids particle size	79
solid treatment technologies	134–135
solid waste contactors	164–168
solid waste management	7
solubility	79
solvent extraction	130
special waste	3, 26–31
special waste consignment	
note procedure	195–196
specific gravity	79
spillages	159
spills and leaks	69
spot or grab sampling	78
stabilization	124
stacks	157
statutory	
instruments	21
nuisance	34, 43–44
undertaker (SU)	45
steam stripping	130
storage, interim	16
storage tanks, underground	77
stream assessment	70

strict liability	56	VOCs (see volatile organic compounds)	
supervisory system	51	volatile organic compounds	
surface treating	69	(VOCs)	178
Sustainable Development	7	control of	112
systems approach	104	volume reduction	124

T

targets	97–98	
techniques of waste minimization	92–97	
technology		
end-of-pipe	101	
changes	94	
Total Organic Carbon (TOC)	80	
Total Suspended Solids (TSS)	79	
toxicity	179, 81	
trade effluent notices	159, 45	
transboundary movements	52	
transfrontier shipments of waste	173–174	
transport		
dangerous goods	168, 84	
waste	168–173, 195	
treatment methods, selection	102–105	
treatment technologies		
gaseous	105–124	
liquid	124–134	
solid	134–135	
Tremcards	170	
trespass	47	
TSS (see Total Suspended Solids)		
turbidity	80	

U

UK Water Industry Act 1991 (WIA)	45
ultrafiltration	130
underground storage tanks	77
unit operations	104
US environmental law	57–61

V

viruses	178

W

WAMITAB (see Waste Management Training and Advisory Board)	
waste	
assessment/survey	65–78
audit	165
carrier	44–45, 167
commercial	26
contractors	18
controlled	3, 26, 37
database	84
definitions of	24–31
disposal	196
generated during a fire	159
hazardous	51
industrial	26
law and regulatory authorities	12–13
licensing system	38
liquid contractors	164–168
manager	11
minimization	14–16
mix	17
packaging	53
regulation	31–44
solid contractors	164–168
special	3, 26–31
streams	13
transfrontier shipment of	173–174
transport of	168–173, 195
waste disposal	3
off-site	18–19
contractors	165
facility	165
plans	51
waste management	4

general principles	7–12	water conservation	92
licensing	38–43	watercourses, discharges to	159
Waste Management Industry Training and Advisory Board (WAMITAB)	42	Water Resources Act 1991 (WRA)	46
		wet air oxidation	124, 128
		WRA (see Water Resources Act 1991)	
waste treatment	3		
on-site	16–18